A Sociology of Energy, Buildings and the Environment

Bringing the social sciences to the heart of the environmental debate, this book demonstrates the relevance of sociological analysis for environmentally critical issues like energy consumption. Focusing on energy efficiency and the built environment, the authors take a critical look at the production and use of technical knowledge and energy-related expertise. Challenging the conventional assumptions of scientists and energy policy-makers, the book outlines a new role for social research and a new paradigm for environmental policy.

Supporting the central argument are three key case studies:

- A history of the insulation industry, illustrating the erratic character of technological innovation.

- A review of housing development, challenging conventional notions of the factors behind standards of energy efficiency.

- An analysis of new office building, throwing new light on the idea that technology transfer is impeded by non-technical barriers.

Drawing upon a wide programme of empirical research, the authors extend the reach of sociology and of energy research and policy. This book, therefore, represents essential reading for sociologists, students of environmental topics, and energy policy-makers and practitioners.

Simon Guy is Reader in Urban Development and Director of the Centre for Urban Technology at the University of Newcastle. He has undertaken sociological research into a wide spectrum of urban design and development issues funded by the Economic and Social Research Council, the Engineering and Physical Science Research Council and the European Union.

Elizabeth Shove is Director of the Centre for Science Studies at the University of Lancaster. She has undertaken research projects relating to the construction industry, energy and the environment, everyday technologies, ordinary consumption and research, and science policy.

Routledge Research Global Environmental Change Series

British Environmental Policy and Europe
Edited by Philip Lowe

The Politics of Sustainable Development
Edited by Susan Baker, Maria Kousis, Dick Richardson and Stephen Young

Argument in the Greenhouse
Mick Mahey, Stephen Hall, Clare Smith and Sujata Gupta

Environmentalism and the Mass Media
Graham Chapman, Keval Kumar, Caroline Fraser and Ivor Gaber

Environmental Change in Southeast Asia
Edited by Michael Parnwell and Raymond Bryant

The Politics of Climate Change
Edited by Timothy O'Riordan and Jill Jagger

Population and Food
Tim Dyson

The Environment and International Relations
Edited by John Vogler and Mark Imber

Global Warming and Energy Demand
Edited by Terry Barker, Paul Ekins and Nick Johnstone

Social Theory and the Global Environment
Michael Redclift and Ted Benton

A Sociology of Energy, Buildings and the Environment

Constructing knowledge, designing practice

Simon Guy and Elizabeth Shove

London and New York

First published 2000
by Routledge
11 New Fetter Lane, London EC4P 4EE

Simultaneously published in the USA and Canada
by Routledge
29 West 35th Street, New York, NY 10001

Routledge is an imprint of the Taylor & Francis Group

© 2000 Simon Guy and Elizabeth Shove

Typeset in Garamond by The Midlands Book Typesetting Company,
Leicestershire.
Printed and bound in Great Britain by Biddles Ltd, Guildford and
King's Lynn.

British Library Cataloguing-in-Publication Data
A catalogue record for this book is available from the British Library

Library of Congress Cataloging-in-Publication Data
Guy, Simon.
    A sociology of energy, buildings, and the environment :
constructing knowledge, designing practice / Simon Guy
and Elizabeth Shove.
        p. cm.
    Includes bibliographical references and index.
    ISBN 0-415-18269-7 (alk. paper)
    1. Energy conservation--Social aspects. 2. Energy conservation--
Environmental aspects. I. Shove, Elizabeth, 1959- II. Title.

TJ163.3 .G79 2000
333.79′16--dc21

                                                                00-058256

ISBN 0-415-18269-7

This book is dedicated to the late Edward Joseph Guy, to Beryl Guy and to John and Jocelyn Shove

Contents

Figures and Table

Acknowledgements

It is hard to know where this story begins. For Elizabeth Shove it probably starts with a day trip to a technical workshop on energy efficiency and buildings. This was totally new territory for a sociologist and not one that promised to be very interesting. By way of comfort, Stuart Sutcliffe, then assistant director of York University's Institute of Advanced Architectural Studies, explained that it should be an intriguing day, even for a sociologist, since energy was an issue that linked all the design professions together. Simon Guy's was a baptism of fire. Fresh from cultural studies, he plunged into a project, the first step of which was to interview people in the insulation industry. Following energy efficiency ever since we have strayed through uncharted territory, confronted new puzzles, and met a huge variety of practitioners, policy makers and researchers who have puzzled with us.

The Economic and Social Research Council has a lot to answer for in funding sociological research into energy efficiency and buildings and we would like to thank them for Award No. L320253021. We would also like to acknowledge the support of our sociological friends and colleagues, many of whom are still bemused by our enthusiasm for the worlds of energy efficiency, design and construction. Equally we want to recognize our new friends and colleagues in building science, architecture, engineering, and property development, many of whom are still not sure quite what it is we do. We hope this book provides some explanation.

1 Introduction

James Thurber's grandmother 'lived the latter years of her life in the horrible suspicion that electricity was dripping invisibly all over the house' (Thurber 1963: 68). The idea that electricity might leak from empty light sockets is bizarre but at the same time strangely plausible. Both everywhere and nowhere, energy remains an often mysterious feature of everyday life. For many people, hazy memories of physics lessons and dimly recalled descriptions of electrons flowing this way and that confirm the sense that energy management is a largely *technical* issue. It comes as no surprise, then, that it is technical rather than social experts who have provided estimates of remaining fossil fuel reserves and who have been responsible for assessing relationships between energy consumption, carbon dioxide emissions, and global warming.

Although anchored in science and technology, energy-related practices and policies have swung sometimes dramatically in response to the changing status of energy consumption as a 'problem'. The oil crisis of the 1970s turned energy use into a national political and economic issue, creating questions to be addressed and resolved where none existed before. The focus during this period was on finding ways and means of rapidly decreasing dependence on oil. This immediate pressure revived longer term anxiety about the depletion of non-renewable resources, bringing back fears and forecasts about when fossil fuel based societies might reach the limit of their reserves. Since then, the nature of the energy problem has shifted subtly but significantly. Roughly half of all carbon dioxide emissions relate to the energy consumed in buildings. Re-cast as a critical environmental concern during the 1980s and 1990s, the need for energy conservation acquired a new significance.

Whether framed in terms of resource depletion, national security or global environmental change, the task of reducing energy consumption has generally been defined as a scientific rather than a social challenge. This is especially so when the focus is on the built environment. Sticking with the case of buildings – as we do throughout this book – the problem has been seen as one of reducing the heat lost through walls, roofs and floors and of increasing the efficiency of systems for heating, cooling and lighting. This has in turn generated a range of questions for building research, and an associated array of theories about the process of technology transfer and the relationship between science and practice. Such theories and expectations themselves depend on an implicit understanding of the social world and of the conditions and contexts of change in building construction and building use.

Written by a pair of sociologists, this book is about energy-related knowledge and practices in the professional worlds of building research, design and construction. Energy policy makers and building researchers appear to share a remarkably uniform view of change and how it comes about. Yet it is a view that is often at odds with the every day experiences of those who construct and inhabit the built environment. In following these themes and tensions, first examining the construction of the energy 'problem' and then unearthing associated theories of social change that underpin research and practice in this field, we develop a distinctive strain of environmental sociology.

We argue that it is distinctive because of the rather pragmatic approach we take to the problem of explaining the social dynamics of energy consumption and understanding the social organization of energy-conserving measures and practices. We believe that sociological concepts and theories can make a really important contribution to environmental debate and policy, but as we outline below, we also suggest that some of these opportunities have been missed. This is partly because the discipline has been preoccupied with its own theoretical concerns and partly because sociologists have been drawn into, and have sometimes accepted, ready-made but limited roles addressing the so-called 'human dimensions' of energy efficiency.

Environmental sociology and energy efficiency

For sociologists, as for many others, 'the environment' has represented a new arena in which to extend existing theories and across which to replay long-standing controversies. Sociologists of social movements have, for instance, turned their attention to the development of environmental beliefs and values. Likewise, debates about the relationship between nature and the social world have been traced back to Marx, Weber and Durkheim and extended forward to encompass current environmental concerns (Martell 1994; Redclift and Woodgate 1994; Bell 1998). Much the same is true for analyses of risk and uncertainty or the science of climate change (Beck 1992; Lash *et al.* 1995; Shackely and Wynne 1996). In other words, 'the environment' has been appropriated by the discipline in ways that reflect prior dispositions and theoretical orientations. It is no wonder, then, that social constructivists battle against realist opponents (Hannigan 1995; Burningham and Cooper 1998), or that social environmental researchers disagree amongst themselves about questions of methodology and approach. As 'the environment' has been absorbed into the discipline, so it has been incorporated by different theoretical perspectives.

Not all environmental themes have proved to be suitable subjects for sociological treatment. Despite their global environmental significance, and in accounting for half of all carbon dioxide emissions, energy use in buildings really is significant, mundane issues of everyday energy consumption lack the qualities of visibility, public interest and theoretical appeal that characterize other more fashionable debates.

We begin this book, a book that is unashamedly about the sociology of energy, by reflecting on how a sociology of energy and the built environment might relate to more conventional environmental discourse. From time to time, social theorists have seen energy as a key variable in cultural evolution. As Loren Lutzenhiser observes, such positions typically begin by recognizing that:

> societies require energy-conversion technologies to survive, that the amount of available energy to a culture is a function of technology, that energy-conversion differentials influence the relative prosperity and power of societies (limiting in some cases and stimulating dramatic growth in others), that technology choices are determined primarily by political processes, and that polity and economy are in turn shaped by cultural institutions (e.g. religious, scientific, governmental and corporate arrangements, understanding and belief).
>
> (Lutzenhiser 1992: 54)

Periods of human history have therefore been inspected and explained in terms of the interplay between natural resources, the technologies of their exploitation, and attendant social systems. Events such as the transition from coal to electricity, or the introduction of the steam engine surely had far reaching societal consequences so it is no wonder that they have been subject to social-historical scrutiny (for example, Cottrell 1955; Nye 1998)

Although these sorts of macro-sociological interpretations have been influential in their own way and important in adding a materialist foundation to social history and theory, they tend to invoke somewhat functional explanations. It is as if social organization is 'driven' by the marrying of basic human 'needs' (for food, shelter, etc.) with available resources. Contemporary investment in identifying and cata-loguing national and international flows of energy and resources (see, for example, research at the Wuppertal Institute[1]) represent environ-mentally inspired variants of earlier efforts to reveal trends and inequalities in energy production and consumption (Redclift 1996). The work of the Inter-governmental Panel on Climate Change is a good example of such an approach. Again the scale is big and again the ambition is to map and understand determining factors. Concep-tualizing the economics and politics of natural resource management on a global scale is clearly significant, and especially so in the case of energy (Grubb 1991; Giovannini and Baranzini 1998). Yet recogni-tion that societies are to some extent shaped by the energy technolo-gies on which they depend, or that there are massive inequalities in the distribution of energy resources worldwide, does not necessarily help when it comes to conceptualizing energy as an environmental problem, or thinking through the implications of established institu-tions and policy paradigms for the production and incorporation of energy efficient knowledge and technologies.

Moving one step closer to everyday practice, energy researchers are increasingly interested in the relationship between energy use and the ordering and organization of everyday life. There is, for instance, some evidence to suggest that people's expectations of comfort are increasing (Wilhite *et al.* 1996; Cooper 1998); and that the energy intensity of taken-for-granted routines and habits is steadily rising. However, energy consumption has managed to retain a remarkably low profile as an issue of public environmental interest. Numerous surveys and studies have revealed extensive public 'ignorance' of the link between energy use and carbon dioxide emissions, and equally impressive misunderstandings of where and how energy is consumed within the home (Hedges 1991; Löfstedt 1992; Farhar 1993; Hinch-cliffe 1995; Kempton *et al.* 1995).

Striking images associating boiling kettles with climate change and the ravages of hurricane damage made a simple visual connection between everyday practice, energy use and global climate change.[2] Yet these links have yet to catch the imagination of consumers, campaigners or policy makers. On the face of it, this is curious because scientific assessment of the part that energy consumption plays in contributing to global climate change should be enough to secure its position high on the public as well as the official environmental agenda (Department of the Environment 1993). Certainly it should be enough to fix it squarely on the agenda of environmental sociologists. So what is the problem?

In its day-to-day invisibility, energy has much in common with other subjects that have nonetheless become high profile topics of acknowledged environmental importance. Levels of biodiversity are just as difficult to detect with the naked eye and there is a sense in which it requires as much of an act of faith to believe in the 'reality' of energy consumption or to 'see' the energy embodied in an aluminium window frame as it does to believe in a hole in the ozone layer (Yearley 1991: 120). With energy, as with biodiversity, knowledge is made and mediated through modelling and measurement and in each case, scientific expertise is on hand to document and explain what is going on. Perhaps it is the sheer familiarity of energy use, and its deep embeddedness in taken for granted patterns of everyday life, which makes it so especially elusive.

Maybe it is the lack of any tangible symbol. Popular environmental concerns are much more commonly organized around a handful of icons: polluting chimneys, pandas, whales, and even ozone layers offer a focus for attention that it is hard to match. Crusades against nuclear power apart, it has proven difficult to turn energy, and especially routine energy use, into an emotive subject around which to organize and campaign. Those who advocate simple, environmentally friendly lifestyles still rely on electricity, and there is no doubting the vested interests at stake in maintaining not just an industry, but a network of physical infrastructure and a corresponding array of modern Western expectations of a normal life, all of which presume an extensive and complex energy system.

For some of these reasons, and for the time being, energy consumption fails to make the grade as an item of popular concern and, as such, it fails to make an appearance in sociological analyses of such anxieties and the movements they have spawned. In short, the relative insignificance of energy as a topic of public interest excludes it from direct consideration by many environmental non-governmental organizations, and by those interested in social environmental movements (Beck 1992, 1995; Eder 1993).

But perhaps there is more to it than that. In trying to explain why sociology paid such little attention to the environment in the late 1980s, and in trying to understand what he describes as 'a collective failure of the sociological imagination' (Newby 1991: 6), Newby introduces a further argument, suggesting that the long-standing challenge of distinguishing the social from the natural represents a form of disciplinary boundary work: by marking out a distinctively social realm, researchers also mark out a distinctively sociological role for themselves. Hence their reluctance to get too involved in issues that challenge this boundary, or that cross the line and address questions about the natural world and its physical resources. Extending this point, Lutzenhiser notes 'Sociology's own theoretical unease with technology and the physical/natural world, and its insular tendencies in regard to other disciplines' (Lutzenhiser 1994: 58).

No wonder, then, that energy consumption and conservation so rarely figures as a subject of sociological analysis. Not only is energy invisible and energy use routine, mundane and seemingly essential, it is also tied up with the technicalities of engineering and design. The treatment or, more accurately, the marginalization of energy use in environmental sociology is symptomatic of all these tendencies.

Before abandoning the search for points of connection and contact between current sociological debate, on the one hand, and environmentally critical but routine issues of energy consumption on the other, we should check for signs of relevant activity in the sociology of science and technology. There are two reasons for looking in this direction. One is that energy consumption is intricately linked to science and technology on a number of counts: with respect to the proliferation of devices and gadgets that use energy; to types and forms of energy production and, in the longer term, to the replacement of human labour with energy extracted from non-renewable resources. More significant still, ways of life have co-evolved with the energy-consuming technologies on which they now depend. We might therefore expect to uncover analyses of energy consumption disguised as social histories of science, or cropping up in the course of debate about states, industries, regulation and the development of new technologies.

Sure enough there are a number of explicitly energy-related studies in this vein. For instance, Hughes (1983) examines the weaving together of institutions, interests and national electrical infrastructure. Cowan (1983) addresses similar issues, this time looking at the co-evolution of domestic gadgets and domestic labour within the household. Closing in on just one product, Bijker (1992) uses the case of the fluorescent light to explore and challenge conventional arguments about technological diffusion. Akrich also draws on energy technologies, taking the development and export, but not use, of photo-voltaic kit as a case with which to unpack received wisdom about 'appropriate' technology (1992). Taken together, this work (and, in particular, ideas about the inter-dependence of socio-technical change, the 'lock-in' of technology, and the way in which present and future technologies organize or 'configure' their users) draws attention to the structuring of opportunity and the parts played by 'non-human actors' including existing infrastructures and technologies as well as embedded practices and routines. These examples are relevant not only in that they are in some sense directly about energy, but also because of the potential and the promise they hold for re-thinking the relationship between the social and the technical.

Looking back, it is easy to understand why the energy efficiency of the built environment has failed to attract more sociological attention. Energy is invisible, building design is a technological process, there are no obvious theoretical footholds and, perhaps most important of all, there are many other more amenable environmental issues on which to concentrate.

Engaging with energy and buildings

We take an alternative approach. Rather than simply thinking about the opportunities that environmental issues generate *for* sociology, and thus evaluating environmental topics in terms of their appeal within the discipline, we consider the kinds of contributions sociologists could make *to* environmental debates. An initial aim of the Economic and Social Research Council's Global Environmental Change Programme was to 'take the social sciences to the heart of environmental debate and to take environmental debate to the heart of the social sciences': an aim that also points to the possibility of bringing sociological imaginations to bear on such novel themes as acid rain, waste disposal, and even energy efficiency. Much of the research on which this book is based was supported by that programme and informed, if not inspired, by the notion that sociology has something to offer when it comes to understanding environmental issues and developing environmental policy.

If we take natural scientific and policy assessments of environmental priorities at face value (at least for the moment) and suggest that sociologists can and perhaps should engage with the issues these raise, a new agenda begins to appear. From this perspective, issues are important not because of their potential relevance back home – back,

that is within the discipline – but because of their perceived significance in terms of global environmental change. Swivelling social environmental research priorities around in this way, the 'problem' of
energy use in buildings becomes a central rather than a marginal
concern (Lovins 1992).

The sort of sociology we want to develop in this book represents a
deliberate attempt to engage with a specific area of environmental
research and policy. As such it represents a kind of experiment, illustrating both the problems and the potential of trying to take a particular brand of social science to the heart of just one environmentally
significant sector. Put another way, we are not primarily interested in
exploring energy use in buildings because this allows us to contribute
to theoretical debate within environmental sociology or the sociology
of science and technology. Instead, we set out to see how sociological
methods and arguments drawn from these fields might be applied and
how they might fare alongside more conventional, more technological,
representations of the energy problem.

In thinking about possible forms of sociological engagement and in
contemplating the sort of contribution sociologists might make with
respect to energy efficiency in buildings, it is important to acknowledge that the field is not entirely open. Technologists and energy
policy analysts already have quite clear views about social science and
the types of questions on which such expertise might be brought to
bear. Social scientists wandering into this territory are therefore
confronted by ready-made roles relating to what are usually described
as the 'human dimensions' of energy efficiency (Shove and Wilhite
1999).

It is widely recognized by policy analysts and technical researchers
alike that the availability of technology does not ensure its use. This,
together with a growing sense that reduction of carbon dioxide emissions is a really urgent environmental imperative, has helped to focus
attention on 'non-technical factors' believed to govern energy
consumption and limit the realization of proven technical potential.
Social and market researchers are consequently commissioned to
undertake studies designed to improve the efficacy of energy-saving
campaigns, assess their impact, or inform predictions of future energy
demand (Hirst and Brown 1990; Hedges 1991; Hirst 1992).

In coming to the rescue, and in attempting to answer questions
framed in this way – why won't people save energy, what are the
barriers to energy efficiency, how can they be overcome, and so on –
sociologists tacitly accept their sponsors' two-part view of a world in
which proven energy-saving technologies confront a reluctant social

world. This may well be a world in which social scientists have a valued role as expert guides, but in accepting such a function they run the risk of undermining the potential for more systemic forms of social analysis.

Rather than joining forces with those who adopt the role of the 'people experts', we have tried to steer another course, starting in a different place in order to tackle different questions. Instead of jumping into the space already reserved for social science we want to take a few steps back and reflect on the theories of change and action inscribed in energy-related building research and policy. How is the energy problem conceptualized? How is technical research expected to relate to energy-efficient building practice? What are the tacit understandings of relevant knowledge and appropriate practice? By turning our attention to these more macro issues we create an agenda for the sociologies of science, technology and the environment. The exercise that follows is thus one of unearthing and sometimes challenging theories of change and action that routinely underpin energy research and policy.

Science, knowledge and practice

We begin by looking at the evolution of the energy problem from the point of view of research managers and others involved in funding and promoting building science. Given the variety of national research systems, not to mention differences of building practice, culture and climate, we perhaps naively expected to discover correspondingly diverse forms of energy expertise. Instead, we show how standardized methodologies and agendas foster the development of a relatively uniform epistemic community of energy researchers. This population of experts deals in necessarily abstract knowledge, and in the production and management of theories and models that transcend the one-off challenges faced by individual building designers. As we show, the world of funded research is organized around the manufacturing of 'cosmopolitan technical regimes' (Disco *et al.* 1992, Rip 1992) and the production of energy-related knowledge abstract enough to circulate through publications and international conferences, and standardized enough to constitute agreed advice and guidance.

It is one thing to know how to build a low energy office, but another to be in a position to actually do so. Related theories of technology transfer implicitly support the relevance and legitimacy of generalized research and development. The problem of implementation is consequently positioned, fair and square, as a problem for individual practitioners. However, our three case studies suggest that opportunities for adopting energy-saving strategies are anything but standardized, individualized or economically determined. They show that technologies and energy-related practices are selectively appropriated within specific social contexts. These accounts of practice make it clear that similar technical strategies do and do not make sense for different reasons and at different moments in time, and that their adoption depends on the sometimes competing perspectives and priorities of a whole network of organizational actors. Whatever else, the picture is certainly not one in which proven knowledge is seamlessly transferred from research to practice.

Nor is it one that is unchanging or in which the economic and environmental imperatives of energy conservation stay still. As the case studies also show, certain energy-saving strategies and technologies are easily accommodated, positively welcomed and actively promoted within the industry. For example, techniques for producing and installing mineral fibre insulation have advanced dramatically over the last forty years or so. Research that succeeds in improving the quality of insulation is likely to have wide ranging impact partly because such developments make relatively little difference to builders, designers or occupants. Once fitted, insulation can be forgotten. By comparison, other energy-conserving measures have consequences for the design of the building as a whole, and for how it is used and inhabited. For example, effective implementation of the 'principles' of passive solar design requires an elaborate process of case-by-case interpretation, taking account of the orientation, layout and the materials of which each building is made. Generalized guidance about passive solar design is therefore subject to multiple layers of translation as it filters into specific, always localized, design decisions.

As these examples remind us, building research and building science operate across different levels of abstraction and generate a range of variously bounded insights and conclusions. More than that, the processes of translation and interpretation required to produce more energy-efficient buildings are themselves socially situated. These observations point to two tensions both buried within this discussion and both running through the rest of the book. The first concerns the nature of building knowledge and the relationship between apparently transferable insights from building science and the seemingly unique characteristics of building production. The second thread concerns the relationship between technical and social situations. Energy-related advice is so wrapped up in a technical framing of the problem that it is expected to be relevant and applicable across a wide range of social contexts. Yet our case studies show that energy-related practices are socially specific and localized in terms of time and context.

Although this book is in a sense about the production and application of building science, it explores tensions between social situations and technical expertise, and between cosmopolitan and local knowledge in terms that are themselves generalizable. In this way, our discussion of the built environment engages with questions that are central not just to the promotion of energy efficiency, but to other areas of environmental policy as well.

Sociological analyses of high profile environmental controversies for instance surrounding genetically modified organisms, the disposal of radioactive waste or the management of water pollution also address issues of knowledge, policy and practice. Yet the themes that preoccupy scholars working in these fields, for instance, issues of public trust, or the global politics of environmental agenda setting, are not especially salient when the focus is on the ordinary technologies of energy efficiency or the publicly invisible process of building construction. Although we have appropriated established concepts of knowledge, technology, and social change, we have had to make some modifications, sometimes re-tooling entirely in order to make sense of events within this particularly inconspicuous realm of environmental policy. In searching for relevant resources we have been drawn towards the sociology of science and technology. Ideas about the 'locking in' of new technologies, or about moments of irreversibility in the history of design prove to be useful tools for the analysis of individual devices and objects. Yet, with buildings, we confront not one object, not even one standardized system, but a unique assembly of many component parts. In order to conceptualize design processes and comprehend the interaction of shifting populations of manufacturers, suppliers, occupiers, developers, builders and professional experts, we have had to adapt and extend existing methods and concepts.

Despite this mixed approach, some features are clear. Although written by sociologists this book is not going to help readers searching for an understanding of the 'human dimension' or for advice on how to overcome non-technical barriers to energy efficiency. It is not about anticipated responses to price signals or regulatory strategies, nor is it about individual choices and people's environmental values, attitudes and beliefs. We say nothing about how to design information materials or energy campaigns for maximum impact. Although these are words of warning, they also signal an invitation to other social scientists and make a challenge to policy makers and to technical research and design communities.

Our proposition is that the social sciences can make a much greater contribution to energy and environmental policy than is normally supposed. Rather than falling into a conventional advisory role, we take a more critical stance, using the case of energy efficiency and buildings to reflect on the theories and models of social change that sustain environmental research and policy. This exercise allows us to identify and explore alternative ways of viewing social and technical change, and to reflect on the roles of policy makers and other actors implicated in such processes. Read in this way, the book has much to say about how policy perspectives and initiatives rest on a bed of tacit but nonetheless questionable ideas regarding choice and action, agency and structure.

We begin this experiment in social science by examining the various research environments in which energy-related knowledge is defined and produced. Chapter 2 shows how the energy problem has been defined and who has been involved in addressing it. Puzzling over the uniformity of building science, Chapter 3 shows how shared conventions and methods of technical enquiry inform and are in turn supported by a dominant techno-economic model of research, development and practice. Taking this a stage further, Chapter 4 explores the usually implicit theories of social and technical change that sustain programmes of energy research, and introduces alternative approaches drawn from the wider sociological literature. Making use of these insights, Chapters 5, 6 and 7 investigate various contexts of energy-saving action, focusing on the fate of insulation in four countries; on efforts to improve the energy efficiency of the same building type (housing) in different organizational settings; and following the dynamics of energy-related decision-making within the world of commercial office development. The final chapter considers the theoretical and methodological implications of a more sociological approach to the analysis of energy efficiency. We argue that such a shift of perspective promises to generate more realistic and more useful accounts of the relationship between knowledge and practice and of the social processes ordering this and other environmental issues.

2 Building research environments

The meaning of energy efficiency and its place within a wider reper-toire of research and policy priorities is far from stable. Recognition that fossil fuel reserves were finite, and early acknowledgement of environmental risks meant that energy had a place on the building research agenda prior to the 1970s oil crisis. Yet it was the combina-tion of suddenly escalating prices and political anxiety about national security and dependence on imported oil that really put energy on the map. This move also involved a re-definition of the 'problem'. Long-term concern about the depletion of fossil fuel reserves gave way to the short-term pressures of price and politics.

A brief history of the slogans used in UK government energy-saving campaigns gives as good a picture as any of the evolution of the energy problem over the last thirty years. The 'Save It' campaign of the 1970s concentrated on conservation: it was about turning off lights, lowering thermostats and cutting back on demand. The next wave, introduced by the injunction, 'Get more for your Monergy', was about the promotion of efficiency, not conservation *per se*. Rather than cutting back, the goal was to maintain expected levels of service (for example, in terms of comfort or lighting), but to do so using more efficient systems and technologies. The third phase introduced a totally new vocabulary. The posters and leaflets produced by the UK Energy Efficiency Office in the 1990s carried the headline, 'Saving the Earth Begins at Home', and featured illustrations of children dancing around a spinning globe. Global warming and the need to reduce carbon dioxide emissions provided a new rationale for energy research and prompted a new round of technical and scientific activity. As in the 1970s, the emphasis was on conservation but this time for environmental rather than political or financial reasons.

These re-definitions of the purpose and importance of energy consumption have filtered through the world of building science, and even a brief history of research and development programmes would show how energy-related priorities have been translated into new knowledge, methods and techniques and hence into the design of the built environment. This chapter, which is as much about research and science policy as about the substance of energy research, follows the re-definition and management of 'the energy problem' and its incorporation into technical research agendas. How was the need for energy conservation or energy efficiency conceptualized as a research problem, what new knowledge was thought to be required in the field of building science, and how has such knowledge been developed and translated into practice in the UK, in Sweden, Ireland, the USA, France and Finland? These were the questions we pursued with the help of the sixty government officials, building scientists, and technical researchers we interviewed, and through an analysis of technical literature, project reports and research reviews.

Standing some distance back, we detected a number of rather predictable patterns. In all the countries we studied, funding for energy research followed a similar course, increasing dramatically in the 1970s, dipping as energy prices fell in the 1980s and gradually picking up in response to the environmental concerns of the 1990s. A bird's-eye view of research priorities shows both a 'maturing' of the field and its adaptation to new pressures and interpretations. The first 'phase', especially in the UK, was dominated by an interest in quantification and measurement: how was energy actually used and what were the most appropriate targets for conservation? Diagrams showing the proportion of energy 'escaping' through walls, floor, roof and windows were all the rage and featured prominently in text books and guidance documents of the late 1970s and early 1980s. The evidence of these input–output models helped establish priorities for further research.

Initiatives to develop new materials and more efficient heating and cooling systems followed, as did an interest in the thermal performance of buildings as a whole. Understanding the potential for passive solar design, that is for buildings that deliberately made better use of the sun's energy, was fashionable for a while, as were complex exercises in modelling building performance. The environmental turn inspired new research directions, for instance prompting studies of embodied energy, that is the energy consumed in making building materials such as aluminium, insulation, and so on. Life cycle assessments, whole-life costings, investigations of the potential for recycling building materials, and of designing for durability have extended the range of energy and environmental research,[3] but have not radically altered its framing in terms of physical properties and material characteristics.

At first sight, this is not an especially surprising narrative. The first step was to measure the problem, the second to develop solutions. The third challenge has been to ensure that the results of research and development are adopted in practice. As we discuss in Chapter 4, efforts to understand the barriers and obstacles that seem to impede the uptake of proven technology have almost as long a history as technical research itself. For the time being, this brief representation of energy-related research and development gives a sense of the key concerns and their evolution over time.

Defining energy research

Before considering the development of energy agendas in more detail we reflect on the general characteristics of the trajectory sketched above. What sorts of questions were included and which fell outside the remit of enquiry? As we have already noted, the first step was one of measurement. Early efforts to quantify the flow of energy within buildings allowed researchers to estimate the proportions 'lost' through various elements of the structure. But this was not the only sort of measurement that might have been undertaken. Later studies of recorded energy consumption in technically identical dwellings showed that one household might be using between two or even ten times as much energy as a neighbouring family (Lutzenhiser 1997; Socolow 1978; Diamond 1984). Rather than constructing a research agenda on the basis of actual energy consumption, as recorded in 'real' households, the technical community focused on the building alone. This seems to be a strange move. After all, it is not as if buildings are autonomous beings that literally use energy. On the other hand, this purifying strategy has a number of important advantages for energy researchers. Given the diversity of human practice, it is reasonable to argue that the physical properties of energy performance can only be studied systematically once buildings have been actually or theoretically emptied of their unruly occupants. For these and maybe other reasons, the course of building research has been shaped by the quantification of input–output flows rather than the analysis of actual energy usage.

The narrative of building research has also been affected by the cost of energy. Most obviously, resources for energy research have kept pace, inversely, with the rise and fall of energy prices. More obliquely, judgements about the relative merits of alternative technological strategies have been, and still are, founded on forms of cost–benefit analysis, estimates of pay-back period and the like. In this way, assumptions about economic rationality, choice and action have been embedded in the design of technological research programmes and in the evaluation of relevant questions and appropriate solutions. Reliance on standardized methods of cost–benefit analysis effectively obscures potentially important cultural variation in the meaning and value of energy. What is missing from this framing of building research is an understanding of what energy consumption means to householders or building occupants, of where their priorities lie, and of how they value the services which energy makes possible. In short, energy-related building research has been dominated by a view of the building as an essentially physical entity with uniformly physical and technical properties.

The ability to abstract the problems of building science in this manner is particularly impressive given the diversity of construction practice. To give just one example, the details of building construction in southern France are quite unlike those routinely adopted in Finland. Given the relationship between indoor and outdoor climates, not to mention the vastly different cultures and traditions of construction, one might expect to see highly localized programmes of technical enquiry. As later sections of the chapter show, there are variations in how the 'energy problem' has been translated into a problem for building science. However, it seems that these differences have more to do with the organization and funding of building research than with specific histories of design and practice.

In order to position itself as a branch of science, energy-related building research has discounted the otherwise confusing practices of building occupants, and reduced and then incorporated economic considerations through standardized methods of cost–benefit analysis. Further complexities, in the form of existing conventions of construction practice, are located outside the margins of relevant enquiry, but are given an important part to play with respect to subsequent 'marketing' initiatives and programmes of technology transfer. Developing this argument later in the book, we suggest that a building science that seeks universal applicability by bounding its subject in this manner is likely to run into trouble. For now it is enough to note that energy-related building research has been inspired by such a model of science and that this partly explains why technical research agendas follow a broadly similar course despite technically relevant differences of building design and construction, and despite localized, historically flexible, interpretations of 'the energy problem'.

Studies of science policy and science funding tend to focus on more glamorous themes: on resources for high-energy physics, the funding of very expensive facilities, support for biotechnology, and so on (Mulkay 1979; Cozzens *et al.* 1989). In cases such as these, the flow of resources and the specification of research priorities have a tale to tell about the societal negotiation of knowledge, mediated through the institutions of research and science funding. The sheer scale of support is clearly relevant, but so are the details of research management. How are research questions established, how are agendas formed, what are the institutions of science, and what part do funders and researchers play in this process?

Asking similar questions of the more ordinary territory of building research is just as instructive. A comparative review of systems of research management and methods of research funding allows us to unearth anticipations and expectations surrounding the production of specialist knowledge of energy efficiency, and helps explain the otherwise puzzling convergence of research priorities. In the building sphere, most energy research and development is of an applied nature. The goal is to produce methods and measures that are relevant to practice and fit for application. This feature is important for it means that the processes of priority setting are doubly revealing: not only do they tell us about prevailing interpretations of 'the energy problem', they also contain clues about current practice and how research is expected to influence it.

Comparing research environments

Comparison of the organization and management of energy-related building research programmes in France, the UK, Finland, Ireland, Sweden and the USA allows us to investigate the generation of technical priorities. From the mid 1970s onwards, the common goal has been to reduce energy consumption in the built environment. How has this been handled in countries that have different histories of building, different economic and political positions regarding energy supply, different climatic conditions, and different traditions of government funded research? In addressing these questions we have limited ourselves to a review of research and development activity directly supported by government. This is a restrictive feature for it means that we only ever capture a part of the wider research scene. Second, we have tried to provide a picture of what are, in practice, constantly evolving research environments. Some of the specific arrangements we describe have already been superseded or overtaken by events. This is important for the cases involved, but does not compromise our analysis of four generic research situations.

Before describing these in more detail we should say something about how we followed the formation of research programmes in the six countries included in our study. Having identified managers of relevant government funded research and development programmes we began with a simple question: 'How are priorities determined?' In almost all cases, respondents resorted to a model of technical potential. Their shared ambition was to concentrate research effort on those areas where there was greatest potential for energy saving, and where proven measures might be widely replicated (this is, of course, to assume that technical potential can and will be realized in practice). In an ideal world, research proposals would be judged, selected and specified accordingly. Although this is a reasonable ambition, assessment of technical opportunity assumes extensive knowledge of the nation's building stock, and of the possibilities and problems that represents. Although lacking the data required to construct a blueprint for future research, managers nonetheless subscribed to the view that priorities should be determined with reference to this abstract concept of technical potential.

How was this impossible yardstick made real? When asked again how priorities were determined, research managers documented an array of administrative procedures, all designed to provide approximate if not perfect technical insight. Some relied upon the collective wisdom of an advisory committee. Others identified the most rewarding lines of research and development through analysis of aggregate data and computer-based estimates of the building stock.[4] As we explain below, strategies varied depending on the local research environment. Yet the general point remains. The ideal of setting priorities with reference to a map of technical potential was an important part of the research culture, even for those who recognized this to be an inevitably elusive goal.

In terms of practicalities, research managers were right to describe administrative procedures when questioned about the setting of research priorities. These systems determined the blend of interests and expertise on which judgements of value and potential were founded. In the sections that follow, we review systems of research management in terms of the relative influence of government, the research community and the construction industry. Second, we reflect on the relationship between building science and building practice: how is knowledge developed in one domain expected to flow into the other? Finally, we take note of the limits of energy-related building science: where are the boundaries set and what falls beyond the scope of government funded research and development?

The six countries included in the study were initially selected in order to explore the differential effects of climate, culture, and building practice. In the event, the structure of the research community and the prevailing system of research management was of much greater significance for the formation of energy-research programmes. The following discussion therefore distinguishes between four research environments: *close communities*, characteristic of countries with a small population; *co-ordinated contractors*, in which research funders interact with sizeable research groups; *contracting knowledge systems* in which both research and project management are fragmented; and *networks of expertise*, in which research funders forge alliances between research providers and users. These types are not mutually exclusive. Nor do they exist in isolation. Research systems are not homogenous and just as patterns and practices change over time, so different regimes co-exist within individual countries. In describing these four types, the point is not to characterize whole nations (even if certain regimes dominate), but to explore the implications of each type for the definition, production and promotion of energy-related building research.

Close communities

Sweden, Finland and Ireland have relatively small populations of building researchers and although it is not literally true, respondents in all three countries shared the feeling that 'everybody knows everybody else'. Research and development programmes are devised by government departments or agencies [The Irish Energy Centre (formerly EOLAS) in Ireland, NUTEK and the Swedish Council for Building Research in Sweden, the Ministry of Trade and Industry in Finland], in collaboration with representatives of industry and in conjunction with researchers typically based in university departments or government-funded research institutes. This is not an especially unusual arrangement.

What is distinctive is the extent to which the parties involved have experiences and histories in common. Government officers, researchers, and key players in the building industry had often taken the same courses at the same universities, hardly surprising since only one or two institutions offered appropriate technical training. With fewer people involved, occupational divisions were not as rigid as elsewhere and we encountered instances in which people switched between industry, government and academia as their careers developed. One

Swedish respondent, currently working in the insulation industry, had been seconded to the government to help develop new building regulations; others regularly moved between the roles of university researcher or teacher, government advisor, and private sector consultant. Advisory boards and research programme committees inevitably included a number of familiar figures, each with varied and overlapping experience and each likely to have a practical as well as a professional interest in defining research directions. Although competition for funding was 'normal' the relatively limited number of centres involved had a tendency to specialize and to dominate the field in specific areas.

As respondents readily acknowledged, this comfortable intimacy had benefits and limitations. National research initiatives had to grapple with the double pressure of delivering relevant results whilst also providing researchers with longer term career opportunities. Small research communities cannot expect to develop a full range of specialist knowledge, and interaction with other countries was important. In the three cases we considered, considerable effort was invested in developing such links. For example, the Swedish government supports the International Energy Agency's Centre for the Analysis and Dissemination of Demonstrated Energy Technology Analysis Support Unit (which has close links with Chalmers University of Technology in Gothenburg), partly as a means of building expertise and developing an international reputation in this field. The Irish Energy Centre, a joint initiative of the Irish Department of Transport, Energy and Communications and Forbairt (the national agency responsible for technology and business enterprise) is itself supported by the European Union through the community support framework. In Ireland, as in Sweden and Finland, participation in the European Commission's THERMIE and JOULE programmes has technical as well as financial advantages. Despite or rather because of their close-knit characteristics, these research environments were especially open to international influence and exchange, a theme we take up again in Chapter 3.

The small country setting had further implications for the relationship between research and practice. Research managers in Ireland, Finland and Sweden had a relatively intimate knowledge of the national building stock and of major energy-related initiatives. In Sweden the 'Stockholm project', which involved the construction and monitoring of some 200 apartments between 1983 and 1985 was, for instance, extremely well known. While such major schemes are unlikely to be repeated, respondents from different backgrounds shared a common knowledge of the people involved in this and subsequent initiatives and competitions. Although no longer part of their brief, government experts in Ireland recalled how they used to provide advice and guidance on a one-to-one basis for individual householders. In these cosy worlds, the immediacy of first hand experience frequently blurred distinctions between the technical and the social.

In these contexts, discussions about the technical potential for energy conservation were consequently imbued with tacit understanding of local socio-economic possibilities and socio-cultural characteristics. In addition, the small scale of the research community seemed to foster interdisciplinary exchange or at least reduce the significance of disciplinary boundaries. Swedish representatives of government, industry, and academia independently explained that they saw themselves to be part of a long-term collaborative research effort. Their accounts might be rather too glowing, but in this apparently trusting environment, project managers, researchers and, representatives of industry claimed to respect and draw upon one another's technical expertise. Certainly there was little gap between research and practice. Indeed there were instances in which industry members of advisory boards literally *were* the market for the research in question.

The *close community* setting affects the forming and framing of technical research but does not determine the detailed substance of energy conservation. Swedish experts were, for example, increasingly preoccupied with questions of indoor air quality and sick building syndrome. A six-year research programme on 'the healthy building' (started in 1997) illustrates the extent of these concerns and the scale of the funding on offer (Abel *et al.* 1999). In these respects, Ireland faces a very different set of technical problems having only begun to regulate thermal insulation in 1992. Indeed there is a sense in which the location of technical trouble and the focus of technical research

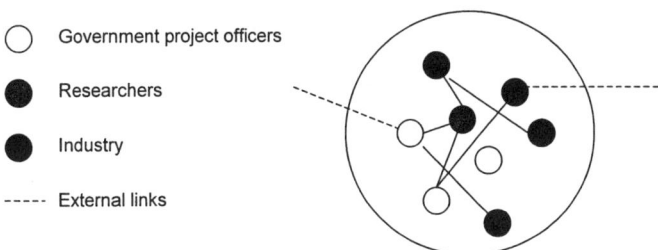

○ Government project officers

● Researchers

● Industry

----- External links

Figure 1 Close communities

shifts with each adjustment to 'normal' standards of energy efficiency. Past patterns of investment in the built environment have equally telling consequences. Swedish tax incentives, combined with building regulation, have fostered comparatively high standards in terms of space, specification and energy conservation. The relatively widespread use of district heating in Finland and Sweden also presumes a certain socio-economic environment, a particular form of local government and a specific set of beliefs about house, home and local community.[5] Such factors clearly influence the list of items on the agenda, but it is the research environment that shapes the way in which those questions are conceptualized and addressed.

The diagram above represents features of the *close community* in graphic form. As the image suggests, those involved in the research system are held together by a circle of mutual inter-dependence.

To summarize, the *close community* permits interaction across disciplinary and sectoral boundaries. Research agendas mirror the interests of a small, relatively stable, group of people whose backgrounds reflect the concerns of the building industry, of materials producers and manufacturers, the priorities of government, and the commitments of social and technical research communities. Problems are viewed in all their complexity for it is difficult to separate the socio-economic from the technical at this close range.

Co-ordinated contractors

The *close community* depends upon informal interaction between members of a relatively tight knit group of energy researchers, their 'users' and their sponsors. By contrast, the second arrangement, which we refer to as one of *co-ordinated contractors,* depends upon formalized relationships between research funders and those who do the work. This structure is one in which a limited number of experienced research groups compete for projects commissioned by a central core of government officers. As adopted in the USA, this is a system that allows relevant government departments to develop and fund national research programmes, notwithstanding the fantastically varied contexts of building practice.

The contrast between this environment and the homogenous culture of the *close community* is striking. In the USA, the challenge has been one of devising worthwhile research programmes relevant and appropriate to a vast and unknowable population whose energy-related actions depend upon any number of situationally and climatically specific considerations.

Decisions about research priorities depend not on first-hand knowledge and gut feeling but on a steady stream of aggregate data and a substantial bedrock of statistical analysis. Data on the state of the building stock and the market penetration of selected energy-saving technologies create a statistical portrait of the nation to which federal level research responds. Research contractors and departments commissioning research accept the reality of this 'data-base nation' that they have helped to create. Such simplified abstraction is, it seems, a necessary step in transforming endless variety into a manageable map of technical targets and shared strategic goals.

In the USA, the range of potential research providers is enormous, including consultancies, universities, non-governmental organizations such as the American Council for an Energy Efficient Economy (ACEEE), state level experts, utilities and so on. At first, the national laboratories (for example, Lawrence Berkeley Laboratory or Oak Ridge National Laboratory) represented just one of several sources of research expertise. Unlike other research providers, the national laboratories had a symbiotic relationship with the federal government. There were long established links with the Department of Energy and it was easy to set up, and renew, relatively long-term, relatively open-ended contracts. As budgets shrank during the 1980s, the Department of Energy relied ever more heavily on the laboratory system (Lutzenhiser and Shove 1999).

Although geographically spread, researchers based in the national laboratories constitute a fairly homogenous community, linked to one another and to their project officers at the Department of Energy through their mutual roles in shaping and responding to the federal agenda and in defining programmes of energy research at one (or may be more) steps removed from the localized practicalities of building construction.

Under this arrangement, research directions reflect federal priorities. The recent history of energy-related building research in the USA has undoubtedly been marked, sometimes negatively,[6] sometimes positively, by changing interests in national security, national economic performance, energy efficiency and, most recently, environmental protection (Schwartz 1996). Within this framework, researchers have been able to make proposals and influence their sponsors. Interpretations of relevant research therefore reflect the histories, characteristics and priorities of a limited population of *co-ordinated contractors*. In the case of the USA, three features are particularly important.

First, national laboratory staff have traditionally focused on the production of technical knowledge, paying relatively little regard to the practicalities of its diffusion or suitability. When questions about consumers, users, or industry do arise, the dominant techno-economic orientation effectively pushes them to the margins of relevant enquiry.[7] Second, the national laboratories have tended to specialize. Although there is increasing competition between them, there is also an uneven distribution of expertise. In this context, it is easier to commission laboratory specific studies of discrete technologies than to bring experts together or develop projects that cut across self-defined specialisms – for instance spanning wall technology and window construction, or relating high-tech work on cooling systems to research into passive solar design. One consequence is that relatively little attention has been paid to buildings as a whole.

Finally, there is no place for understanding the routine practices of the building industry, for coming to terms with the habits of designers and users, or for appreciating the varied commercial concerns of companies involved in producing and selling building materials. This is to be expected, for the approach outlined here embodies a classic model of research, development and dissemination. Positioned in this way, government has a right, maybe even a responsibility, to focus upon basic and strategic research, technology transfer, and the analysis of market barriers. From that point onward, individual companies and industries are expected to take over. Incentives and initiatives such as the 'Golden Carrot', 'Green Light', and 'Energy Star' programmes are designed to include industry in the promotion of energy efficiency but without straying too far from the 'proper' remit of government involvement.

Technology transfer has nonetheless been an important item in the Department of Energy's research agenda. Over the last twenty years, laboratory researchers and others have learned about marketing energy-related information. They have tracked the diffusion of new techniques and investigated and occasionally quantified the merits of alternative promotional strategies (Warkov and Meyer 1993; Macey and Brown 1990; Brown and White 1992; Vine *et al.* 1993; Vine 1995). In what remains a largely technical research environment, such work concentrates on the fate of proven solutions and strategies once they reach the market: hence the emphasis on individual decision-makers and on the reactions and responses of what are taken to be rational economic actors. Other potentially significant questions, for instance, regarding the cultural significance of energy services or people's changing expectations of comfort, lie beyond the margins of normal enquiry.

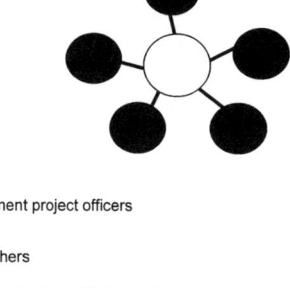

Government project officers

Researchers

Formal contracts and informal ties

Figure 2 Co-ordinated contractors

To summarize, systems of co-ordinated contractors revolve around a central agency working in collaboration with a limited number of research providers. As Figure 2 suggests, roles and responsibilities are formally defined, yet there is scope for informal interaction when shaping priorities. In this context, research providers' interests inform perceptions of relevant and appropriate lines of enquiry. Where these interests are largely technical, there is only limited scope for reaching beyond the confines of the dominant paradigm. This is a typically bounded system, perhaps even more restricted in terms of scope and range than the close communities described above. Interaction with researchers outside the immediate circle is limited and agendas grow by accretion, one project leading on to the next.

Contracting knowledge

Like the system of *co-ordinated contractors*, this method of research management depends upon a clear distinction between those who undertake research and those who decide what new knowledge is required. The difference is that projects are typically shorter and more precisely defined, and the list of potential contractors is generally longer. In practice these minor details have important consequences for the development of technical knowledge.

In illustrating the characteristics of this arrangement we refer to UK's Building Research Establishment's Energy Conservation Support Unit (BRECSU). The Building Research Establishment was privatized in 1997 (Courtney 1997) and no longer manages government-funded energy-related building research in quite the same way. Though it has now joined the ranks of research providers, competing alongside universities and other research institutes for government funded projects or for the role of project or programme manager, BRECSU was until recently responsible for commissioning research on behalf of the government's Energy Efficiency Office.

During this period, research plans devised by individual project officers were submitted to advisory committees and evaluated in terms of their 'replication potential', that is in terms of the technical potential for repeating proven results across the nation's building stock. Anticipated economic and environmental benefits were also quantified in pounds' worth of associated energy savings and tonnes of carbon dioxide. Project officers specified discrete packages of work, which together constituted an agreed programme. Competitive tenders were invited for each package. Anxieties about value for money and accountability were such that project officers retained tight control over the research process. In order to avoid delays or contract extensions, project officers were themselves responsible for describing detailed work plans, defining methodologies, determining the sequence of events and laying down the time allowed for each activity. Research contractors had little or no chance to influence project specifications or to inform the overall direction of the research agenda. The tendering system was designed to ensure short-term cost effectiveness and to promote competition between potential contractors.

Never sure of when or where the next project might come from, research contractors found it hard to accumulate knowledge, develop areas of specialist expertise or retain research teams. It was equally difficult to build alliances between research groups, all of whom were in competition with one another. Not only that, project officers rarely informed researchers about parallel contracts nor about how their project fitted into the wider scheme of things. This style of piece-by-piece management preserved the sense of a truly competitive environment and positioned researchers not as collaborators or co-producers of knowledge, but as contractors and providers of research services. Research agendas therefore reflected the views, beliefs and technical backgrounds of project officers. It was they who devised research strategies and they who determined the detailed content of associated research contracts.

Within this context, project officers sought to construct a coherent programme from a handful of short-term contracts. For a time at least, the managers we studied were so overwhelmed by the administrative labour involved in running so many separate projects that they had little time to piece together the results of these divided labours. To make matters worse, relatively high levels of staff turnover meant

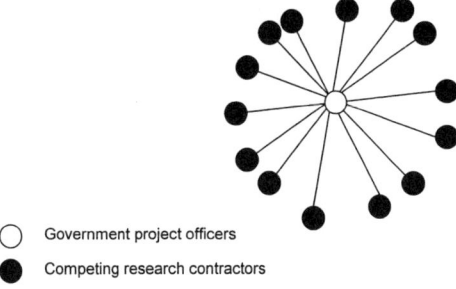

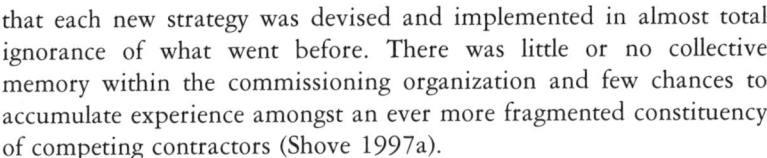

○ Government project officers
● Competing research contractors

Figure 3 Contracting knowledge

that each new strategy was devised and implemented in almost total ignorance of what went before. There was little or no collective memory within the commissioning organization and few chances to accumulate experience amongst an ever more fragmented constituency of competing contractors (Shove 1997a).

To summarize, the system of contracting knowledge involves a very precise division of labour and a clear, hierarchical, specification of roles. As indicated in Figure 3, project officers define research briefs. Organizations competing for the resulting contracts have little influence over the scope of the research or the way in which it is undertaken. This exact packaging of responsibility limits the potential for inter-disciplinary interaction and for the cumulative development of technical expertise. In other respects the system of *contracting knowledge* has much in common with the pattern of *co-ordinated contractors*. The linear language of research, development and dissemination again guides the design and development of deliberate programmes of technology transfer, and again such programmes aim to overcome market barriers believed to impede the otherwise inevitable uptake of proven energy-efficient technologies.

Networking expertise

The last research environment to be considered depends upon and positively encourages collaboration and interaction between contractors and between these groups and agencies involved in commissioning research. In this context, research agendas emerge through a process of negotiation involving regional as well as national agencies and a network of established research teams. Of the countries included in the study, France provided the best example of this arrangement. At the time of the research, regional delegations provided half the funding for projects sponsored by the Agence de l'Environment et de la Maîtrise de l'Energie (ADEME). As a result, the national research programme was strongly influenced by regional needs and by regionally specific problem-solving priorities. In addition, research frequently required the co-operation and collaboration of other funding bodies, sometimes other ministries, sometimes industry partners as well. Patterns of co-financing were such that groups and agencies which routinely commissioned research also undertook projects on their own account and competed for funding from other sources. Disciplinary boundaries and sectoral divisions were correspondingly complex: so much so that one respondent suggested that it would be quite impossible for any outsider to understand the roles and interests of all the government agencies with a stake in energy-related building research. In the midst of all this, ADEME saw itself as 'an interdisciplinary skills centre and an interface between specialist networks in France and at the international level' (ADEME: 1993).

Although the networks of relevant experts were complex, the simple call for tender remained a critical instrument. Within ADEME, project officers and advisory committees outlined areas of work and invited responses from interested parties. This system has some similarities with the method of contracting knowledge described above. The difference is that invitations to tender are designed to foster positive interaction between the many players involved in energy research. Contracts were typically large enough to sustain several researchers for a year or more. There were annual calls for tender, and projects were described in broad terms, allowing research teams to determine their own research design and methodology.

As one project officer explained, research funding is used to generate 'healthy' differences of perspective and understanding. Although there are recognizable themes and strands of current concern, these are informed by a process of dialogue involving government, industry and academia. This method of research management takes account of existing interests and networks and uses these to foster the development of robust, socially viable technologies. Given the problem-solving focus and the inter-disciplinary character of many project teams there is a parallel tendency to explore issues from multiple perspectives. Accordingly, there have been opportunities to consider the energy efficiency of whole buildings and to include explicitly sociological analysis of the beliefs, practices and preferences of building users (Dard 1986).

As active members of the total research network, social researchers were not necessarily obliged to study the 'human dimensions' of energy efficiency, nor were they required to provide the 'human stuffing' for greedy computer models. There has been scope, and even funding, to examine the nature of innovation and to look again at theories of technology transfer. ADEME has, for instance, supported studies of the social organization of engineering and the management of building professional knowledge, and has funded research on the dynamics of 'techno-economic networks' (Callon *et al.* 1992).

Such work, and such engagement with the realities of research management, has helped generate a range of extremely influential tools and concepts within the field of science and technology studies. It has also raised important questions about the role of government agencies within the field of innovation, and in particular with respect to the promotion of energy-related research and development.

If, as is argued by Callon *et al.* (1992), technical change is mediated by relationships between those involved in the design, production and use of buildings, the effective promotion of energy efficiency depends upon careful understanding of present patterns of interaction. Having identified points at which this social system is 'weak', and having recognized threadbare patches in the network of necessary inter-relationships, the challenge is that of mending and modifying those interactions. While this may involve the production and dissemination of yet more technical information this is not always the case. This analysis re-positions the justification for government involvement. Rather than being restricted to the roles of sponsoring research and development and investing in programmes and measures to overcome market barriers, government has a legitimate part to play as facilitator and 'network manager'.

To summarize, this approach to research management develops a range of institutional positions and intellectual perspectives. As represented in Figure 4, this is a system that requires and fosters

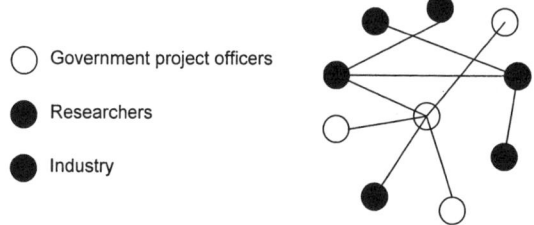

○ Government project officers

● Researchers

● Industry

Figure 4 Networking expertise

Table 1 Comparing research environments

	Close community	Co-ordinated contractor	Contracting knowledge	Networking expertise
Influence of government, research, and industry	Agendas are informally negotiated: all three interest groups are involved	Research responds to policy but can also shape the agenda: industry is not directly involved	Research is commissioned by government project officers: industry is not directly involved	Government project officers formally set the agenda but do so in collaboration with industry and the research community
Relationship between building science and building practice	The two communities intersect	Building science is remote from industry	Building science is remote from industry; but under increasing pressure to be 'useful'	Formal co-operation and informal networking creates space for collaboration and interaction
Positioning the energy problem	Disciplinary boundaries are relatively flexible: the technical is tinged with social understanding	The priorities of a largely technical research community dominate	The priorities of a largely technical population of government project officers dominate	The extensive network of interests involved permits a 'socio-technical' understanding of 'the problem'

communication between the various groups involved: between regional and national agencies and between universities, utilities, industry and private research centres. The resulting network of expertise is central to the way in which specific problems are defined and addressed and to the conceptualization of socio-technical change.

Positioning energy efficiency

These four simplified and sometimes exaggerated sketches show that energy-related building research has involved different players in different settings and countries, and that the details of research management influence the bounding of agendas and the specification of goals, ambitions and priorities. The mix of public and private sector participation varies, as do relationships between those commissioning and undertaking research. Each system of research management

generates a distinctive balance of responsibility and a unique configuration of demands and expectations.

Drawing these threads together, Table 1 summarizes the characteristics of all four research environments in terms of the relative influence of government, the research community and industry; the
relationship between building science and building practice; and the
positioning of 'the energy problem' in terms of social and technical
expertise.

This review suggests that the micro-politics of knowledge production make a real difference to the interpretation of energy efficiency as
a problem for building research. Given that energy-related building
research is designed to influence practice, it is extremely important to
note that the links between research and practice, and hence strategies
and ideas about how to cross the divide between them, are configured
differently in each environment.

There are, nonetheless, points of commonality. Although levels of
formality vary, researchers and representatives of industry, including
the practitioners who are expected to benefit from the results of
research, regularly interact in situations where communities are close
or where expertise is deliberately networked. The interests of practitioners and users are not directly represented where knowledge is
contracted or where contractors are co-ordinated. This is significant
for the active involvement of 'industry' representatives weakens the
grip of more technologically oriented communities of researchers and
government project officers. Such involvement creates the space for,
or perhaps even requires, a rather broader interpretation of relevant
research.

A further theme, and one that we take up in Chapter 4, concerns
the conceptualization of technology transfer. Energy research and
development is of little practical value unless it makes a difference to
those who design, operate or inhabit buildings. Understandings of
how research relates to practice, from the formation of relevant questions to the promotion of subsequent insights and results, also differ
in each of the four research environments. But again there are points
of similarity. Systems of contracting knowledge and co-ordinating
contractors depend upon a clear distinction between knowledge
producers (i.e. the technical experts) and consumers (i.e. users and
practitioners). In rather different ways, the other two arrangements
encourage a blurring of boundaries, sometimes to the point at which
the production of new knowledge is explicitly seen as a process of co-
production.

None of this suggests the wholesale development of new institutions.

Neither have we seen the proliferation of adaptable, inter-disciplinary, problem-oriented research suppliers locked together in what Gibbons *et al.* (1994) refer to as a 'Mode 2' system of knowledge production. We have, however, identified a range of co-existing styles of research management, some of which are more rigidly bounded, more flexible or more permeable than others. We have also argued that these arrangements influence definitions of relevant energy-related building research.

At this point it is important to acknowledge the dynamics of research management. Although we have illustrated the four types with reference to experiences and practices observed in different countries, we do not suggest that these types are fixed for all time, or that they are strongly associated with individual nation states. Taking the case of the UK as an example, the following section shows how approaches to technical research have changed during the last 30 years. Some of the trends identified in this brief historical account can be seen elsewhere. For instance, there are signs of an increasing emphasis on the relevance of research and the role of users in the USA. Likewise, interpretations of the proper role of government are on the move. We have not yet seen a wholesale re-definition of government as 'honest broker' or a recognition that 'traditional structures for knowledge production are not satisfactory' (Gibbons *et al.* 1994: 152), but there is a growing sense that something more is needed if energy research is to make a difference on the scale now required.

Changing research environments

Contributors to a special issue of *Building Research and Information*, on 'the future of national building research organizations' (Seaden 1997) lamented the loss of what was once an extensive, well-funded network of government sponsored building research institutes, common in many European countries, and in New Zealand, Australia and Canada (Shove 1998a). These institutes, like the national laboratories in the USA, have been home to a cadre of specialist but above all independent building researchers. Free from the pressures of the commercial world and from the demands of a teaching programme, institute experts generated a bedrock of knowledge on which their respective industries could draw. Referring back to this 'golden age', contributors and commentators bemoaned the impact first of privatization, and second of increasing industry involvement in research, arguing that both compromised what had been an immensely valuable institutional form.

A similar story might be told with respect to energy research in the UK. In the post-war years, the Building Research Station at Watford (now the Building Research Establishment) was the focal point for construction research, most of which was undertaken by directly employed researchers of various disciplines (Lea 1971). The first major programmes of energy-related research reflected an enthusiasm for problem-oriented scientific enquiry and multidisciplinary team work. Yet they were also marked by developments in science policy as represented in the Rothschild report (Command 4814 1971). Rothschild introduced the principle that government-funded research and development should be undertaken on a customer–contractor basis. This separation of the role of customer and provider led to the dismantling of in-house research, and the creation of a new breed of project managers.

During the 1970s and early 1980s, contracts were large enough to keep researchers going for years at a time. The system was effectively one of *co-ordinated contractors* in which established research groups, mostly but not exclusively drawn from the university sector, were able to exploit a 'gravy train' of funding thus retaining teams of increasingly experienced staff. Within this broadly collaborative environment, researchers and project managers were able to re-define priorities as programmes developed. Through the late 1980s, pressure for greater accountability led to the sharpening of management procedures. As described above, projects were divided into smaller and smaller portions, thereby increasing competition between contractors. The focus on value for money and competitive tendering had sometimes unintended consequences for customers and providers, and for the form and scale of interdisciplinary research.

Although considerable resources were invested in getting the messages of research across, the contracting knowledge regime was not especially well attuned to the changing interests and capacities of potential research users. The UK government's White Paper, 'Realizing Our Potential' (Command 2250 1993) has been extremely influential in making the case for greater 'user' involvement and reconfiguring research policy to that end. This, together with the privatization of the Building Research Establishment, has had the effect of encouraging, if not forcing, new ways of thinking about the kind of expertise that is 'really' needed beyond the perimeter fence of the research institute itself.

The clock does not stop and, looking forward, we might speculate on the possible implications of these trends. One scenario is that further privatization of project management will lead to even closer contract control and a narrowing of the energy research agenda for administrative and managerial, if not intellectual, reasons. In this case we might expect the grip of the dominant techno-economic paradigm to tighten even further. This would be bad news for researchers dealing in what seems to be uncertain knowledge about the complex and often unpredictable processes of energy efficient design and practice. Such a development would also influence perceptions of relevant and important areas of enquiry, and would in turn restrict the government's capacity to act as an 'intelligent customer'.

A more positive scenario is one in which reference to 'users' brings with it a requirement to re-engage with questions about the social as well as technical or economic viability of energy-saving technologies. This might involve a whole hearted revival of long buried agendas relating to the patterns and practices of energy use. Analysis of the conditions and circumstances of socio-technical change would then become central, not marginal issues. The simple question, 'Who are the real users of current technical research?' threatens to undermine established positions and priorities and open the way for new lines of enquiry.

In conclusion, this chapter has documented alternative regimes of research management and commented on the way they change over time. The four research environments we have considered generate and delimit more and less likely lines of enquiry, and more and less open interpretations of the 'energy problem'. Looking to the future, we identified two generic trends: one involving the privatization of once publicly funded research activity, the other favouring the greater involvement of potential research users. The implications are uncertain. Such moves might have the common consequence of broadening what constitutes relevant research, or they might encourage a retreat to a more predictable, more manageably technical agenda. Other developments, including the internationalization of energy-related research, for instance, through funding from the European Union, or with support from the Conseil International du Bâtiment (CIB) are similarly ambivalent: recognition of the localized diversity of energy-saving potential might increase as a result. On the other hand, the goal of exchanging ideas and techniques might lead to their further abstraction from the messy and confusing contexts of action and practice. How these possibilities turn out and how systems of research management evolve in the future, whether they converge around one model, whether they diverge further or transform along the way, will make a difference to the range and type of expertise brought to bear on the question of energy efficiency. But will it also make a difference at the level of research design and methodology or to the fine grained production of knowledge within universities, research institutes, international project teams, and interdisciplinary networks? This question is taken up in the next chapter, which asks what kind of knowledge is knowledge of energy efficiency? This begs the further question, do methods of knowledge production sustain what Haas (1990) and others refer to as an epistemic community? Do building researchers constitute a coherent community that persists regardless of the fluctuating systems of research management that have been the subject of this chapter?

3 Energy knowledges

The previous chapter showed how different research systems influenced definitions of the energy problem and the formation of related programmes of technical enquiry. Research managers invested in projects and programmes that promised to produce applicable results, and to some extent research directions reflected distinctively national agendas. In asking how assessments of technical potential were themselves generated we strayed into the territory of research and science policy. Our comparative review of research environments suggested that the relative influence of government, academia and industry had tangible consequences for the specification of technical problems and priorities. Research strategies differed as a result. It was not simply that academics sought to 'capture' resources to pursue their own scientific interests, or that greater involvement of industry 'users' necessarily transformed interpretations of relevance. Yet the extent to which agenda setting processes either disregarded or were dominated by an understanding of the possibilities and potential for action in the 'real' world had a pervasive influence on the direction and character of technical enquiry. Or so it seemed.

These conclusions are not entirely convincing for our study also revealed a surprising consistency in the methods and techniques adopted. What kinds of knowledge did these shared methodologies generate? Such questions prompted us to pay more attention to the making of building science and the place of scientific knowledge alongside the other types of expertise.

Although familiar in other sectors too, debates about the relationship between scientific and practical knowledge are especially complex with respect to energy. This is in part because energy is itself a contested concept. To know that one has saved energy, one has to have some way of rendering it visible. Knowledge and even definitions of energy depend on the methods of measurement involved and as we show techniques for 'capturing' the dynamics of consumption and conservation differ even within the realm of building research. We therefore begin this chapter with a discussion of how energy is known to researchers and to practitioners and consumers. This provides the starting point for a more detailed analysis of the practice of building science. By looking at what building scientists do we are better able to understand the curious co-existence of a range of research environments and what seems to be a remarkable uniformity in the methods, assumptions, and strategies of building science.

In exploring the theory and practice of building science we show how researchers have isolated energy and separated the science from the building. These moves allow technical experts to create a social and intellectualspace in which to produce universally generalizable results. Like other scientists, building researchers 'tend to assume that natural processes are consistent all over the globe unless there are some very good reasons for not doing so' (Yearley 1996: 101). We suggest that such assumptions, together withan accompanying reper-toire of methodological conventions, constitute the foundations of an epistemic community of energy-related building re-searchers.

As described by Haas:

> An epistemic community is a network of professionals with recog-nized expertise and competence in a particular domain and an authoritative claim to policy relevant knowledge within that domain ... what bonds members of an epistemic community is their shared belief or faith in the verity and the applicability of particular forms of knowledge or specific truths.
>
> (Haas 1992: 3)

In the case of energy research, strands of methodological consensus appear to cut across otherwise significant differences in the organiza-tion and management of building science. The existence of an expert population whose beliefs and advice span the divides of national interests and politics is important for the formation of international environmental policy. As others have argued, 'institutional cultures cutting across national boundaries may provide important opportu-nities for the formulation and implementation of global environmental agreements' (Rayner 1991: 78). But what are the consequences of such a community for the development of knowledge which is only effective once put into practice, and in which practice is a highly loca-lized, highly context specific affair?

We do not argue that the existence of shared scientific conventions is problematic or that it is impossible to reach generalized conclusions on the subject of energy-efficient design. However, we do suggest that the relationship between research and practice deserves much more explicit consideration. With this in mind, we review a number of international efforts to 'enhance the exchange of information on new, cost-effective energy-saving technologies' (CADDET 1998). Initiatives of this kind presuppose that it is both possible and necessary to circulate lessons of energy efficiency abstracted from specific experiences and demonstrations whether adopted in Japan, Sweden or the USA. Our discussion of these and other programmes of technology transfer helps articulate contemporary ideas about the dynamics of knowledge abstraction and appropriation. This leads us to think again about both the process and the philosophy of technology transfer and the relationship between standardized technological measures and the very variable contexts in which they are produced and put into practice.

Concentrating on the production of energy knowledge, this chapter is in four parts. Part one, *constructing conventions,* considers the problem of knowing about energy use. Having argued that methods of measurement effectively constitute their subject, the next two sections review different strategies for acquiring new knowledge about energy efficiency in buildings. Methods of *abstracting knowledge* are illustrated with reference to research on passive solar design. Alternative means of generating *building knowledge* through case studies and monitored demonstrations are then illustrated. The fourth part, *replicating knowledge*, considers current theories and assumptions about technology transfer and the characteristics of building science and building practice.

Constructing conventions

The routine invisibility of energy consumption presents special problems for those interested in its management and conservation. It is impossible to know if you have made energy savings without first finding some means of identifying and quantifying consumption. There are different ways of going about this task and, as might be expected, technical languages of energy measurement differ from those adopted by building designers, users and owners. Willett Kempton's ethnographic work on the 'folk quantification' of energy reveals an impressive range of understandings about how and when energy is used in the home. Kempton and Montgomery (1982) documented an array of technically incorrect but nonetheless plausible beliefs about the function of switches, thermostats and valves. Right or wrong, these ideas informed householders' actions and practices. Subsequent research on the design and interpretation of energy bills demonstrates that the task of simply informing people about patterns of energy usage is persistently challenging (Egan *et al.* 1996; Wilhite and Wilk 1987). Turning to the vocabularies of technical research, energy is represented and made real through such terminologies as kWh (kilowatts per hour); watts per meter square, BTUs (British Thermal Units); U values (total heat transmission coefficient), and so on. These languages of measurement draw on theories of heat transfer and thermodynamics and, like the folk beliefs referred to above, they too influence actions and practices. Studies investigating the place of technical information in the design process suggest that architects tend to think and operate with the help of abstract concepts (Marvin and Mackinder 1985; Kealy 1989a). Engineers may view energy demand as something to be calculated and quantified but from an architectural perspective, energy efficiency may be seen as an integral part of design, inextricably tied up with the form and performance of a total building project.

These perceptions and interpretations are extremely relevant for energy is only known through such measurements, conventions and concepts. The co-existence of so many discourses suggests that for all practical purposes there are multiple types of energy. Of course, the conventional view is that some vocabularies offer *better* access to the reality of energy usage than others. Most experts would claim that James Thurber's grandmother was simply 'wrong' in her assessment of energy usage (she believed it to flow out of empty light sockets) and would have benefited from further, better, information (Thurber 1963). The idea that professional or technical representations provide a more complex, more accurate, picture of what is really going on is rather well established. It is this which justifies any number of educational initiatives, and it is this that legitimates extensive investment in design tools and the production of simplified advice and guidance. Whether the perceived gap is between scientists and designers or between government advisors and householders, the logic is the same. Special effort has to be made in order to translate lessons and messages established in one domain into languages that others understand.

The notion that what counts as energy is determined by how it is known and experienced is not especially contentious. It is perhaps harder to swallow the theoretical and policy implications of the view that energy is most usefully understood as a form of knowledge or to conclude that different methods of measurement generate different types of knowledge and hence construct what are, in effect, different forms of energy (Shove 1997b). The argument that methods of measurement actively constitute their subject implies quite a radical theoretical re-orientation for those seeking to reduce energy demand by changing (or improving) practitioners' knowledge. As noted above, strategies designed to 'get the message across', across, that is, from experts to others, revolve around the notion that transfer is conceptually straightforward and that translations are possible. By looking at *energy as knowledge,* we begin to problematize these ideas. The implications of this approach are developed later in the chapter but for now we will simply adopt the view that energy is effectively constituted by the means through which it is known. What does this mean for the production of building science?

Abstracting knowledge: capturing solar energy

In this section we look at the ways in which building scientists and physicists have sought to understand and maximize the use of passive solar energy in building design. It is an instructive case to take for it highlights many of the tensions involved in developing and applying scientific methods to the 'real world' of construction. There is certainly nothing new about the idea of selecting south-facing sites or of orienting buildings so as to take advantage of solar gain. Such strategies have been a normal part of building practice for many years.

Despite this long history, it requires considerable methodological ingenuity to demonstrate that a building makes good or better than 'normal' use of passive solar energy. The task of conceptualizing let alone quantifying as yet un-tapped reserves of this resource is even more daunting.[8] Although we all 'know' that solar energy is there, we cannot simply refer to a sun meter, nor check the files for last year's solar bill. If they are to make claims about the merits of passive solar design, or to compare the benefits of one passive solar scheme with another, building researchers have to literally capture the sunshine. The most common response has been to devise methods for comparing or simulating comparison of buildings that are equivalent in all respects other than the way they use solar energy. Recognizing the scientific need for a unified approach, the European Commission's DGXII invested heavily in this area. For example, the objectives of PASSYS, the largest passive solar research project in the world, were to develop reliable and affordable procedures for testing the thermal and solar characteristics of building components and to increase confidence in passive solar simulation techniques. This was a demanding exercise occupying sixty researchers from twenty-eight institutes and ten European countries (Wouters and Vandaele 1993). The PASSYS project involved the development of test cells each equipped with an extensive array of monitoring devices and data collection systems. These identical one-room boxes were manufactured in Germany and shipped to various sites across Europe. Once in place, they were used to test solar 'components' (for example, combinations of wall and window constructions) under different climatic conditions. Despite the levels of standardization involved, further effort was required to distil generic conclusions about the effective exploitation of passive solar energy from the volumes of data generated during the course of this project.

Facing a similar problem, researchers working on a study of passive solar house design for the UK's Energy Technology Support Unit (ETSU) went to considerable lengths to isolate and quantify what they referred to as useful solar gain. By useful solar gain they meant that which replaced the need for heating or cooling. Rather than looking at individual components, the ETSU team sought to make sense of the solar qualities of a whole suite of house designs.[9]

Their first step was to analyse, by means of computer simulation, the energy characteristics of what they termed a reference house. This was a 'normal' house featuring normal levels of insulation, a normal layout and a normal distribution of windows etc. The next step was to simulate the energy performance of a thermally upgraded version of this normal house. Equipped with these two models, the researchers generated a number of passive solar versions of the reference and the upgraded reference design. These hypothetical variants included greater areas of south-facing glazing, smaller windows on the north, new layouts and internal arrangements, an additional conservatory or sun-space and so on. Further modelling exercises were undertaken to evaluate the implications of individual measures and their relative benefits should these imaginary properties be planted down in another part of the country, or given a new orientation on the same imaginary site.

Through a process of successive subtraction, computer analysts were able to calculate thermal loss and tease apart the relative contribution of incidental gains (from people, appliances etc.),[10] and 'auxiliary' heating (from boilers and heating systems). What was left had to be solar gain. Further refinement allowed them to distinguish between solar gain in general and 'useful solar gain', by which they meant that which reduced the demands made of the auxiliary heating system. By repeating this procedure for the reference case, the upgrade reference, and its passive solar variants, the aim was to iden- tify and also attribute the benefits of useful solar contribution to specific design decisions.

In this example, as in other more complex studies, solar energy finally makes an appearance through, and only because of, a forest of methodological conventions. Put another way, solar energy is made real and is in a sense constituted through debate and agreement about measurement, distribution, addition and subtraction. To get this far, researchers have to take for granted further layers of methodological procedure. They have to build on theories of thermal performance, on accepted ratings of different materials, on models of how materials interact to influence total building performance, on estimates of indoor and outdoor climate, and so on. Although such conventions are pretty well established there are persistent areas of uncertainty and long-standing points of difference. Competing models have devel- oped as a result, as has a meta level debate about how simulations should be evaluated (Van de Perre 1993). As the authors of the ETSU overview report observe, there is still 'no single method for deter- mining good passive solar design' (Boss *et al.* 1993: 35).

The point is not to criticize this study or others like it, only to underline the levels of methodological convention and the range of expertise required to drag solar energy out into the open. It is only by working their way through a rather specific sequence of organizing abstractions that researchers can 'see' what is happening. Moreover, they can only 'see' through methodological spectacles of their own making.

None of this would come as a surprise to philosophers of science. As Latour and Woolgar's work on the social construction of facts would lead us to expect, solar energy is 'constructed and constituted through micro social phenomena': it is the result of argument, nego- tiation and consensus (Latour and Woolgar 1986: 236). Having estab- lished methods for demonstrating, capturing and working with solar energy, further lines of enquiry are possible. Once assumptions have been fixed and fossilized, computer modelling allows researchers to 'test' the combined effects of different forms of building construction and to thereby separate and analyse the solar contribution. A crucial feature of this work is its subtractive logic combined with an expecta- tion of future aggregation. Physics provides the un-stated reference point as researchers strive to isolate and so understand the determi- nants of building performance.

If they are to work at all, computer models of the type used in the

[handwritten margin notes: "people → too complex" and "modelling"]

ETSU house design studies have to be slimmed down. Elements of 'reality' are deliberately set aside, at least for a while. This makes it possible to isolate solar energy and come to some conclusions about how it might be exploited. However, those conclusions are not necessarily robust or relevant in the 'real world' precisely because of the simplifications involved in their production. It is therefore tempting, and researchers have been tempted, to make their models more complex in the hope of producing a closer approximation to 'reality'. As always there are selections to be made. Which elements of 'reality' should be added back in and which are in any event beyond the reach of modelling? Other dangers lie ahead. The more reality is included, the more 'unreal' the models become. If overloaded with detail they fail to reveal the qualities of passive solar design with any precision.

For some researchers, the idea of incorporating further doses of reality is inherently problematic. Providing one accepts the theoretical and methodological assumptions buried within the solar energy researchers' computer models, and providing one recognizes their limitations, one might reasonably argue that stripped down simulations offer more plausible insights into energy performance than any amount of reference to the real world. Such a view is rather strongly expressed by two contributors to the EC's PASSYS programme who argue that:

> the results obtained from this work have served to demonstrate that it is very difficult to draw general conclusions from real buildings, with their inherent complexities and occupancy interactions. It would appear preferable to work with unoccupied, simple test rooms or cells.
>
> (Gicquel and Cools 1989: 389).

Others seek to test conclusions reached on the basis of simulation and modelling by actually constructing and measuring the performance of what seem to be optimal design solutions. Approached in this way, the development of building scientific knowledge involves an iterative process in which the predictions of science are (or are not) fulfilled out there in the real world of building practice. In his analysis of *Science in Action*, Bruno Latour challenges this myth of an 'out-thereness' which might be employed in the evaluation of scientific prediction, using the experiences of a group of architects, urban designers and energy specialists to make his point. Having finished their calculations these experts had 'a complete paper scale model of the [solar] village' ...'they had rehearsed and discussed every possible configuration with the best engineers in the world . . . [and had settled] . . . on an optimal and original prototype' (Latour 1987: 248). The scheme failed, not because the solar calculations were wrong but because the real world 'out there' (in the form of resistant Cretan villagers) had been excluded from every other kind of calculation and from the networks and considerations of those involved. In other words, reference to a specific application, out there, cannot be used to arbitrate between theories developed within the solar research community. Success or failure out there is indicative not of the veracity of design theory but, rather, of the extent to which other necessary participants, in this case Cretan villagers, have been enlisted in the solar researchers' network.

Whichever way we look at it, building is a difficult case. Even in 'technical' terms, building performance is strongly marked by such idiosyncratic features as location and orientation. In addition, the relative impact of one measure, for instance the addition of further insulation, depends upon the properties and qualities of the rest of the structure. Isolating and thereby understanding the performance of individual elements may yield comparable and generalizable results. However, this does not necessarily help when it comes to anticipating their combined effect, even though that is what counts in practice. A further sometimes confounding problem is that building science is in any event not the only factor that architects and designers take into account. When dealing with the complex problem of a building, they are not simply dealing with the optimization of a set of fixed parameters. To some extent they are involved in creating those parameters in the first place. Not only that, there are always other priorities at stake, like the cost, style, image and appearance of the end result. Rather than sequentially peeling away layers of intervening factors in the hope of getting at the underlying essence of energy performance, another route is to make a concerted effort to understand how buildings perform in real life.

Building knowledge: demonstrations and case studies

As Loughlin Kealy observes, the study of buildings has been 'seen as an essential component in architectural theory and history, . . . as well as . . . providing information on constructional and technical matters' (1989b: 567). From the 1970s onwards, governments and international agencies have funded energy-related demonstration projects. Buildings incorporating new technologies or design measures have been subject to systematic monitoring and analysis and 'written up' in terms that highlight the lessons learned. The UK's Energy Efficiency

Demonstration Programme began in 1978 and has since been followed by the Best Practice Programme. The European Commission has sponsored various demonstration projects and promotional initiatives such as DGXII's 'Project Monitor' (a suite of forty-nine case studies of passive solar projects in Europe) and the 'maxi-brochures' produced for DGXVII by the network of Organizations for Promoting Energy Technologies (OPET). As described below, the International Energy Agency's Centre for the Analysis and Dissemination of Demonstrated Energy Technologies (CADDET) acts as an international clearing house for case studies and monitored demonstration projects.

Before considering case studies as a means of dissemination and technology transfer we first focus on their part in the production of energy-related knowledge. Analyses of real as opposed to simulated buildings have a number of apparent advantages. First, they provide evidence of actual rather than predicted levels of energy use. Second, researchers can talk with occupants and gauge their experiences of comfort and energy consumption. Third, the fact that demonstration projects exist is itself proof that they can be designed, built and inhabited. The depth of detail and the extent of 'realism' is at the same time a source of trouble. Evaluation and extrapolation depends upon comparison between monitored demonstration projects but on what basis? How much should be known about a demonstration project, which aspects should be recorded and compared, and with what qualifications in place? Skirting round this issue, those involved in producing the Project Monitor series for DGXII explain that: 'the significance of Project Monitor is: – that all the buildings have been used for normal purposes and the reactions of users are recorded; that the energy performance has been monitored in use and the analysis is presented in a standard way; and that the whole scheme is described in a consistent and easily accessible form' (Burton *et al.* 1989: 369).

Establishing a protocol for the production of consistent and standardized case descriptions is both vital and contentious. Where buildings have been retrofitted it is possible to make before and after measurements. Impressive though such data might be, for instance, showing 'savings' of thirty or even forty per cent, evidence of this kind has its limits. For instance, it fails to distinguish between the effect of, say, renewing the heating system or of adding insulation: which measure really made the difference and to what extent was it the combination that counted? Although it is clear that less energy is being used, it is not entirely clear why.

Questions about precisely what is being demonstrated are even more troublesome when there is no previous point of reference. What lessons can be learned from the monitoring of new buildings? Case studies typically outline the designers' goals and intentions and then offer recorded data on energy consumption, comfort, users' satisfaction and environmental performance. Tombazis and Preuss' description of an Athens office building adopts just such a standard format (1999). A general description of the site and the purpose of the building is followed by a catalogue of energy features relating to day lighting, artificial lighting, heating and cooling, and ventilation. In theory these good intentions should have positive effects in practice, but how does a demonstration show that this is in fact the case?

In order to make sense of recorded patterns of energy consumption, further methodological moves are required. Benchmarks have to be set (for example outlining ranges of energy consumption per metre square for a particular building type) and standards established before commentators are in a position to conclude that a new building is indeed a 'demonstration' of energy-efficient construction. As these points suggest, the business of studying the energy performance of whole buildings is as replete with convention as the more selective analysis of specific aspects and components.

Once conventions of independent monitoring and description have been established, it is nonetheless possible to construct a protected realm of scientific research. Libraries of carefully vetted and consistently described case study 'data' can be accumulated and subjected to further comparative analysis. CADDET's analysis support unit, based in Sweden, fulfils just such a function. With the help of 'experts from all corners of the world, the technical and economic results of selected demonstration projects on a particular technology are compared' (CADDET 1998). As with experimentation or simulation, methodological standardization makes it possible to discern patterns of energy use and on that basis extract generalized conclusions about the nature of energy-efficient design. Lewis and O'Toole, for example, explain how the DGXII's Project Monitor case studies 'set out to provide architects and other building professionals with firm evidence of the benefits of passive solar design principles' (1990: 25).

The key point is that the production of transferable and comparative 'evidence' depends upon a platform of methodological convention. Although demonstration projects and case studies appear to describe 'reality', they do so in necessarily partial fashion and in a fashion that rests on a raft of assumptions about what is and is not relevant to the task of making energy visible. In this respect demonstrations are no more 'real' than computer simulations or experimental research. In all cases building science depends upon a process of selection, abstraction and purification and in all cases these methods permit the production of what is taken to be generalizable knowledge.

Across the board, the philosophy of building science assumes the existence of laws, principles, and basic relationships waiting to be discovered and exploited in the cause of energy efficiency. The PASSYS test cells were, for instance, designed to quantify the physical properties of components and materials in isolation and in interaction. Computer simulations are used to reveal constant features of the relationships they model. Likewise, demonstration projects are relevant in so far as they illustrate ideas that can be abstracted and applied in other contexts. Although detailed methodologies differed, the building scientists included in our research were committed to the production of generalizable and therefore universally applicable knowledge. By standing well back from the complexities of design and use, they sought to generate science-based understandings of building technology that would, by definition, transcend localized forms of practical knowledge and experience. In the process this epistemic community constructed a world of its own making. Having separated themselves from the vagaries of building practice (through careful use of the tools of abstraction including test rigs, simulations or monitored demonstrations), scientists were free to exchange data, swap computer models, and export techniques around the world. Paradoxically, their distance from practice was itself evidence of the practical applicability and generalized relevance of their work.

If the separation from practice is a necessary move in the constitution of building science, what further steps are needed to translate the results of scientific endeavour back into practice? In the next section on *replicating knowledge* we review respondents' ideas about how research relates to practice and about what can and should be done to ease the process of technology transfer.

Replicating knowledge

As we saw in Chapter 2, scientific research agendas were strongly influenced by notions of technical potential. In funding energy-related research and development, sponsors aimed to support projects that had the potential to make a real difference to designers' decisions and to subsequent levels of energy consumption. In this section we re-visit ideas about replication and technology transfer in the light of the above discussion of building science and its relationship to practice.

Computer simulations, experimental testing and systematically monitored case studies promise to provide governments, designers, and building owners with a better understanding of energy technologies. Such an understanding should allow them to make well-informed decisions, and to use less energy as a result. However compelling or convincing, this line of reasoning glosses over the resources and effort invested in relating the generalized conclusions of building science to the world of practice. More than that, it assumes such translation is in fact possible.

It is at least plausible to suggest that building science and building practice define energy in different ways, that each is locked into a distinctive, relatively water-tight, paradigm and that translation is simply not an option (Kuhn 1962; Hanson 1981). If we view energy as the outcome of theoretical and methodological convention within the realm of building science we should take a similar view with respect to building practice. And if such conventions and traditions – whether of science or practice – have their own rules of description, there can be no simple process of translation from one to another. These ideas offer a new perspective on the role of 'design tools' (that is on rules of thumb and simplified computer models) and demonstration projects as means of dissemination and technology transfer. Further consideration of each allows us to unpack prevailing assumptions about the replication of knowledge.

Design tools

Let us start with what the experts refer to as design tools. Design is a sequential process. Decisions made early on influence what happens next and in the production of any building there are points of no return. It is also iterative. Ideas and schemes are the subject of almost continual revision and refinement. When and where might energy-related building science feed into this activity? There are a number of possibilities. Published information, advice and guidance can, for instance, sit on the shelf until the moment it is required. Alternatively, the results of scientific enquiry may be distilled and embedded in design tools, for example, in computer programs or checklists that structure designers' problems and guide cycles of iterative refinement.

Exactly what is most appropriately distilled, or quite how principles are best represented and presented depends upon which moment of the design process is in focus. From an energy perspective, some of the most critical design decisions are made 'early on' and at a stage when there is very little data to work with. For this and other reasons, there is a perceived need for radically simplified, radically scaled down versions of the highly elaborate computer models that scientists use to identify and compare building performance. In the USA, staff of the Lawrence Berkeley Laboratory have been working on the development of computer programs for analysing energy use in buildings since the mid 1970s. Their first objective was to develop a building energy analysis program that could simulate all building types in all climates. Current efforts focus on providing a user friendly interface and establishing a computer-based 'Building Design Advisor'. The aim is to assist building designers in making initial decisions about building mass and orientation and to provide feedback on day lighting and solar gain (Lawrence Berkeley Laboratory 2000).

In an ideal world, designers would be persuaded to pause and evaluate the energy implications of their ideas before committing themselves to irreversible decisions. Simple checklists, cut-down computer simulations and other design tools promise to structure designers' thoughts, allow them to toy with competing options and offer rapid feedback on the energy merits of each. Of course this suggests that it is in fact possible to simplify to the extent required and still come to the same or similar conclusions (i.e. the right conclusions) as those that would have been reached with the aid of a more realistic or more complex analysis. The hope that intelligent 'front-ends' and user-friendly computer interfaces will automatically bridge the gap between sophisticated simulation tools and building design professionals (Hensen and Hand 1993) arguably represents a form of wishful thinking. Although popular, the idea that levels of accuracy can shunted back and forth along a scale of simplicity and complexity is highly problematic.

This is partly because 'back of the envelope' sketches and models have conventions and paradigmatic qualities of their own. They do not correspond to the final artefact in any one-to-one manner. Furthermore, it is simply not possible to add or remove detail at will in the course of what is a sequential as well as an iterative design process. Design tools may well have the effect of changing practice along the lines advocated by building scientists, but not because they involve the application of science or because they innocently mediate between sophistication and simplicity. If they have such effect it is because they draw designers into a conceptual framework in which energy is given attention alongside (if not over and above) other priorities.

There are two points to make here. First, design tools do not simply translate between the languages of science and practice. Like it or not, they have hidden agendas and qualities of their own. The second point concerns their incorporation into routine decision-making and design. Exactly when are they used and how are the 'results' they generate positioned alongside other forms of evidence and experience?

Fedeski's (1991) review of design aids in UK architectural practice 'suggests that most architects make little use of the design aid already published or of the technical procedures that it supports' (Fedeski 1993: 324). His research into the way design concepts are developed in practice showed that the business of producing buildings has a narrative of its own, that there are clear divisions of labour between principal and assistant staff and that cooperation rests on a mixture of explicit and tacit knowledge. Questions of energy and environmental performance were typically background considerations. As such these issues belonged within a realm of unstated experience and what Fedeski refers to as 'un-voiced' communication. Fedeski argues that design tools did not fit partly because they forced unstated issues out into the open, thereby threatening normal ways of working. Far from being neutral media through which scientific insights flow en-route to practice, design tools have political characteristics of their own. As such they sit, sometimes awkwardly, between the domains of science and design. This awkwardness is a feature of other more familiar mechanisms of technology transfer including the case study and the demonstration project.

Case studies

As well as generating new knowledge, monitored case studies and demonstrations represent important means of dissemination. Their popularity in this role is interesting and puzzling. How is it that the case study, the most singular packaging of knowledge one can imagine, is so frequently used to promote the generalized principles of energy efficiency?

The simple idea that what works once can be made to work again sustains the concepts of 'demonstration' and 'replication'. Appealing to little more than the 'evidence of your eyes', pictures and project descriptions encapsulate the results of painstaking monitoring exercises: exercises that were, paradoxically, designed to generate scientific rather than intuitive forms of evidence. The case study format is extremely familiar, but what is it that readers see when they flick through glossy brochures and project summaries or when they check databases of documented examples? The European Commission's Project Monitor studies included pictures, plans, sections and elevations as well as a brief description of the purpose of the project and the results of the monitoring exercise. Although less detailed, the cases listed in CADDET's web-based catalogues also include pictures together with a summary of key facts organized under standardized headings such as country, building type, total energy consumption and so on. This material represents the results of careful and systematic work, but how might a designer interpret these cases?

If the goal is replication, as it is for CADDET and others,[11] what constitutes the relevant unit of interpretation? Is the hope that future designers will produce replica buildings (and if so how close does the match need to be if it is to achieve the same results as those illustrated in the demonstration?); is it that they will extract the 'idea' of, say, a sun-space and try that out in another country and on a building with another purpose and location (and if so, how robust is the demonstration and what are the limits of its relevance?); or is it that readers might be inspired, in rather general terms, to take note of energy efficiency when approaching their next project? The data included in case study descriptions gives some clue as to what the authors take to be important features. It is perhaps up to the reader to work out what these are but by implication case studies contain genuinely transferable nuggets of knowledge (Macey and Brown 1990).

In asking what demonstration projects really demonstrate and in questioning the type of knowledge extracted from them, we are perhaps falling into the trap of assuming that case studies are important because they disseminate information, technical detail and evidence. It is equally possible that they are used as sources of inspiration rather than data, or that their most important role is in legitimating decisions already made. Covering all possible angles, Kealy argues that case studies should deal with their subject at several levels, 'at a down-to-earth level which will allow architects the knowledge that low energy design is worth getting involved in; at a rational level that shows the connection between theory and realization; at a theoretical level which establishes the support provided by research and experiment; and finally at a level which asserts the excitement of discovery of a genuine new architectural dynamic' (Kealy 1989a: 569). These are worthy ambitions, but the problem remains. Relatively little is known about how energy-related case studies are actually used, how they relate to other forms of knowledge, or how the images and experiences they encapsulate are disentangled and re-assembled in practice.

This brief reflection on design tools and demonstrations has suggested that the business of getting the 'messages' of science through to practice is far from easy. This is to be expected. Any number of other authors have commented on the problems of technology transfer and still more have sought to catalogue the barriers and obstacles that litter the way. Yet the point we are making is different. Discussions of the relationship between science and practice generally take the transferability of scientific knowledge for granted, and further assume that practitioners lag behind because they lack the latest knowledge and expertise or because some non-technical barrier prevents them from putting that knowledge into practice. By contrast, we have suggested the existence of distinctly different knowledge communities, each with their own conventions of evidence, methodology and relevance.

Technical experts and those who fund them have adopted various strategies for improving scientific understanding of energy efficiency: some do experiments, others concentrate on case studies or simulations. They are, however, united by a shared belief in the transferability of the technical knowledge they produce. The common model is one in which basic research leads, through a process of development (applied research) and demonstration (real life trials), on to dissemination (promoting the results of those real life trials) and finally to practical application. In this chapter we have argued that such ideas hold the research community together but at the same time set it apart from the worlds of practice.

We might have told a rather different story had we examined other more localized mechanisms of technology transfer. The proliferation of energy design advisory services is interesting in this respect. As developed in the UK, such schemes put building designers in touch with energy experts and building scientists willing to act in a kind of consultancy role. The idea is that design advisors, familiar with the latest scientific research, will be able to identify relevant technologies and apply generic lessons and principles to a particular design problem. This model of consultancy-cum-technology transfer is interesting for interpretations of relevance are made case by case. As McElroy explains:

> Statements about 'consideration' of local climate or 'making use' of fabric mass may seem self-explanatory, but assessment of local weather conditions and quantification of the impact of fabric mass will almost always be site specific.

> (McElroy 1993: 632).

Design advisors focus not on technology transfer as a goal in its own right, but on the challenge of increasing the energy efficiency of an individual project. Rather than promoting generic knowledge, the idea is to sieve the reservoir of existing expertise in order to select evidence and ideas of use in resolving an immediate problem. We return to the routes through which technical expertise is actively and acquired selectively deployed in Chapters 6 and 7. At this point it is enough to note that governments have sometimes sponsored 'bottom up' approaches to technology transfer (in the form of design advisory services), but only on a small scale and certainly not at the expense of more conventional strategies of mass dissemination.

Converging conventions

Having documented a variety of research environments in Chapter 2, we asked how shared methods of technical enquiry cut through these differences. In this chapter we have shown that researchers and those who fund them, subscribe to a mutually reinforcing package of beliefs about the role of government funded research and development, about the definition of energy efficiency and its analysis as a technical problem, and about the challenges of technology transfer and the difficulties of getting the global messages of energy efficiency through to millions of individual designers and builders.

Along the way we have taken issue with some of the basic tenets of energy-related building research. Rather than taking transferability for granted, we have explored the idea that researchers and practitioners may have their own conventions, definitions and interpretations of relevance and evidence. If so, there can be no simple translation between them. More challenging still, we have suggested that technical researchers inhabit an epistemological domain of their own making, necessarily and deliberately cut off from the confusing diversity of construction practice. This helps make sense of the evident tension between supposedly transferable technical expertise and the culturally specific environments in which such knowledge is produced and applied.

As building scientists know only too well, real effort is involved in isolating the parameters of energy performance and extracting generalized lessons from the analysis or simulation of individual cases. As they are rather less well aware, the process of reversing and re-interpreting generic or 'global' knowledge and of putting it to work with reference to localized, context specific design problems is even more challenging.

For most of this discussion we have distinguished between two knowledge domains: one of building science, the other of building practice. This has allowed us to turn the spotlight first on the conventions of scientific knowledge production and then on the design tools, demonstrations, and advisory schemes through which the latest technologies and techniques are (apparently) fed into practice. Our purpose has been to consider the types of knowledge that building researchers generate and those that grow and evolve in practice. Along the way, we have circled around the problem of when and how it is possible to produce practically relevant but also generalizable building science. The difficulties of doing so perhaps explain why there is so often a gap between what might be done in theory and what happens in practice. But the gap is not only a gap of epistemology. Energy-saving actions do not depend upon knowledge alone. In the next chapter, we return to the theme of technology transfer but from a different perspective. Rather than focusing on the commensurability of the knowledge of research and practice we turn our attention to the theories of action, choice and change that underpin programmes of technology transfer and other initiatives designed to promote energy efficiency in the built environment.

4 Theories of knowledge and practice

Chapters 2 and 3 have shown how the challenge of improving the energy efficiency of buildings has been constructed as a task for building science. Across international boundaries and research cultures, the common aim has been to produce 'more and better knowledge about buildings and how they perform' in order to 'identify promising solutions' to the problem of energy inefficiency (Hutcheon and Handegord 1983: 432). Having considered the framing of research agendas and the methods of building science, this chapter takes another look at the link between research and practice and considers the positioning of problems and solutions within what we refer to as a techno-economic paradigm.

As with other branches of physical science, 'the reactions of materials and other inanimate objects are the legitimate interests of the building scientist' (Hutcheon and Handegord 1983: 4). Like problems of rain penetration, fire safety or structural performance, energy efficiency has been defined as a technical issue, amenable to scientific methods of enquiry. The belief that proven technical solutions are transferable and readily applicable to other technically similar situations is widely shared and, on this basis, building research agendas have been geared toward the scientific resolution of what are taken to be physical problems. As Lutzenhiser points out, energy-management strategies have 'focused almost entirely on the physical characteristics of buildings and appliances' (Lutzenhiser 1993: 248). Vast sums of money have been, and continue to be, invested in research with the aim of developing and realizing the technical potential for energy conservation.[12]

The contribution that building science has made to the stock of knowledge about energy efficiency is substantial. Energy-saving technologies and materials have been successfully developed and manufactured, and energy-efficient buildings have been constructed, tested and widely publicized. The technical knowledge and design techniques needed to construct (certain) zero-energy buildings already exists (Vale and Vale 1997). Moreover, extensive monitoring means that researchers know more than ever before about the energy characteristics of the building stock. Based upon such knowledge, the UK's Energy Efficiency Office believes the potential exists for saving some 20 percent of the energy consumed in Britain through the application of proven technology and cost effective measures (Environment Committee 1993b: 2).

The ambition is clear, but the success of this technological mission ultimately depends on the extent to which technical knowledge and expertise is put into practice. The fact that the achievements of building science have not always been taken-up is made clear by brief reference to the latest English House Condition Survey, which estimated that 'only a quarter of English housing was relatively energy efficient' (Department of Environment 1991: 27). Comparison with other European countries confirms the view that English homes 'are the least efficient' with 'comfort levels the lowest in northern Europe' (Schipper 1987: 545). The conclusions of the UK government's Environment Committee report to the House of Commons express 'deep disappointment' that 'little progress has been made ... on turning a consensus about aims into practical achievements' (Environment Committee 1993a: xiii). Puzzled by the 'inadequate diffusion of apparently cost-effective energy-conserving technologies',[13] Jaffe and Stavins (1994: 92) argue that the boundaries of scientific enquiry must be extended. What seems to be missing is an understanding of the non-technical processes involved. Hence, they suggest that social scientists should have a part to play in explaining the mechanics of technology transfer and the relationship between research and practice. But exactly what role should this be and precisely which questions should they address?

This chapter reviews prevailing ideas about technology transfer and the dynamics of social change. Though most of our examples refer to the UK, the ideas at stake are widely shared across the international energy research and policy community. Drawing on recent research in science and technology studies we unpack and challenge some of the assumptions on which these models depend. This helps to outline the terms of an alternative way of seeing the relationship between science and practice, and sets the scene for the chapters that follow. We begin by describing conventional theories of technology transfer and ideas about the 'non-technical' or social barriers believed to impede the otherwise steady application of proven strategies and solutions.

Theories of technical change

Acknowledgement that building practice has not kept pace with building science has led some to conclude that 'building scientists must appreciate the contribution of the life sciences' (Hutcheon and Handegord 1983: 4). Lee Schipper has, for instance, said that those who see themselves as 'energy analysts have made a mistake ... we have analysed energy. We should have analysed human behaviour' (Charfas 1991: 154). The search for insights into human behaviour has led to what has become a long running alliance between building science and economic theory.

Economists have re-interpreted social processes in terms of a market arena that neatly divides the world into separate, but inter-linked, domains of knowledge and action. Assuming perfect information and a definable logic of utility maximizing rationality, further hypotheses are developed regarding the behaviour of individual house-holds and firms. For example, building users who are properly informed about the costs of energy are expected to make choices that ultimately lead to a more rational use of energy. On this basis, a strand of the European Union's fourth framework research programme investigated the 'rational use of energy'. Confidence in individual decision-makers' capacity to quantify the benefits of energy efficiency and act accordingly is central to what we refer to as the techno-economic view of technology transfer.

As we have already observed, energy is invisible. The ability to treat energy as a manageable commodity depends upon some method of measurement and analysis. Energy auditing, for example, of commer-cial and industrial buildings, promises to address this problem and 'enable the new user to cope with the complex conversion equations and calculation of energy costs per standard unit' (Chadderton 1991: 20). The link to economics is explicit;

> An energy audit of an existing building or a new development is carried out in a similar manner to a financial audit but it is not money that is accounted. All energy use is monitored and regular statements are prepared showing final uses, costs and energy quantities consumed per unit of production or per square metre of floor area as appropriate. Weather data are used to assess the performance of heating systems. Monthly intervals between audits are most practical for building use, and in addition an annual statement can be incorporated into a company's accounts.
>
> (Chadderton 1991: 20–1)

This description suggests that financial and energy-related decisions are comparable, and that business and domestic users act as self-interested, knowledgeable and economically calculative agents when considering energy measures. Business users in particular, 'skilled at marginal cost and future calculations', would 'only seem to need to see the potential competitive economic advantage of innovation to move towards energy efficiency' (Lutzenhiser 1994: 868). Following a similar logic, the development of a visible market for energy efficiency promises to encourage domestic consumers to minimize costs by adopting more efficient ways of meeting their needs for energy services such as heating, cooling or water heating.

Just as the methodology of building science depends on the notion that careful monitoring and scientific control generates transferable technical knowledge, so the logic of economics holds that technological potential will be fulfilled in any market situation that demonstrates an 'intertemporal equilibrium that is Pareto optimal' (Harris 1983: 46). Put simply, the assumption is that economically rational actors, equipped with the necessary technical and economic information, will consistently put proven, cost-effective energy saving measures into practice. Hinchliffe makes a similar point, arguing that 'a rational, profit maximizing man is visualized at work, at home and at play', and that 'the engineering model tends to picture humans as optimally utilizing technologies after the fashion of their creators' (1995: 94). This epistemological coalition of building science and economics has led to a view of technological diffusion that revolves around a faith in the replicability of proven scientific knowledge and a closely related belief in the economic logic of consumer behaviour. Lutzenhiser concludes that:

> As a result, a physical–technical–economic model (PTEM) of consumption dominates energy analysis, particularly in energy demand forecasting and policy planning. The behaviour of the human 'occupants' of buildings is seen as secondary to building thermodynamics and technology efficiencies in the PTEM, which assumes 'typical' consumer patterns of hardware ownership and use.
>
> (Lutzenhiser 1993: 248).

The techno-economic view of energy efficiency suggests that if technical knowledge is rigorously tested and demonstrably proven, and if market forces are undisturbed, then technical diffusion should occur smoothly, with the 'right decisions being taken by millions of individual consumers, both at home and in their place of work' (Jonas

- Decisions are made by relatively autonomous individuals - that is, people are free to make energy efficient decisions if they so choose

- People do not make energy efficient decisions because they do not know how to, and/or because they are not aware of the benefits, and/or because there is a price distortion in the market, and/or a conflict of interests (for instance between landlord and tenant)

- Once individuals are personally convinced, have the necessary information and receive the correct pricing and/or regulatory signals they will adopt proven energy saving technologies

- Technology transfer is more or less inevitable - it is part of scientific progress - but can be speeded up by accelerating the diffusion of knowledge and the correction of market imperfections by selective regulation and/or financial encouragement

Figure 5 Ingredients of the techno-economic model of change

1981: 105). Although the process is implicitly inevitable, there may be some delay as news about the benefit of a new technology spreads and as consumer confidence builds. This line of thinking leads to the view that governments and others can increase the rate at which energy saving measures advance along Rogers' (1962) classic 'S' curve of diffusion[14] by providing authoritative advice and guidance. Figure 5 summarizes the key ingredients of the position described above.

This model sets the scene for understanding both success and failure. Success follows where technologies are proven and consumers are rational. Failure, or significant delay in achieving success, is the result of inadequate technical expertise or irrational consumer behaviour, perhaps relating to market imperfections or a breakdown in the necessary flow of information. In this context, the role of government is clear: it is to support technical research and development and help overcome non-technical barriers or obstacles. This generally involves eradicating ignorance by providing information, and adopting appropriate measures to counter instances of market failure.

Barriers to energy efficiency

Proponents of the techno-economic model recognize the importance of occupant behaviour and the role of economic and other forces in shaping technological innovation. During the 1980s and early 1990s, these ideas, and the evident need to address the application as well as the generation of energy-related expertise, became firmly embedded in the framing of energy-related research and development. Social scientists and marketing experts were invited to help understand and address the 'barriers' to energy-efficient practice. The work of the UK's Building Research Energy Conservation Support Unit was, for example, described to the 1993 Environment Committee on energy efficiency in buildings as being concerned with the technical, social and economic factors of energy use and the integration of energy technologies into buildings. At first sight, this represented a significant departure from what had been an almost exclusively technical agenda. On closer inspection, the difference was not so great for the social sciences were invited to address a rather limited menu of non-technical questions. As noted above, techno-economic explanations of failure assume the existence of social barriers. The need to understand, analyse, and overcome such barriers generated a number of studies each concerned to catalogue the obstacles involved (Hedges 1991). The analysis offered by the UK's Environment Committee is typical.

> The experience of the EEO (Energy Efficiency Office) in the United Kingdom, and of endeavours to promote energy efficiency in other market economies over the past 10 years or so, suggests that the main reasons why the natural operation of the energy and associated services market does not lead to the take-up of all cost-effective energy-efficiency improvements is the existence of three wide ranging market barriers to energy efficiency.
> (Environment Committee 1993b: 2)

These three barriers included:

- *Lack of knowledge and information.* As the Committee observed, many consumers are simply unaware of the level of their energy consumption, its relationship to carbon dioxide emissions, and what they could do to reduce it in a cost-effective manner. This information deficit is apparently worst for domestic consumers and small businesses who lack easy access to expert advice.
- *Capital priority.* Those who invest in energy supply take a long-term

perspective. By contrast consumers often require higher rates of return from capital investment in energy efficiency.

- *Market distortions*. Fiscal measures treat the cost of energy and the cost of energy conservation differently. Energy prices fail to reflect full environmental costs.

Having identified these barriers to action, moves have been made to overcome them through raising awareness, providing better information, providing financial incentives where appropriate, and regulating where necessary (Leach 1991). In this case, as in others, policy instruments are chosen and implemented with the aim of removing or at least lowering barriers whether through regulation, subsidy, or advertising and information campaigns (Grubb 1992). These corrective instruments are not equally attractive, and in the UK continuing faith in the power of the market has meant that financial incentives and regulation have been designed to stimulate it as the main lever of technical change.

In other words, legislation and regulation have only been applied 'where necessary', and financial and other incentives 'where other funds are unavailable' (Department of the Environment 1994a: 2). For the most part, effort has focused on the provision of information and advice to end-consumers on the grounds that this will allow informed individuals to make rational decisions which consequently favour energy efficiency. This has sometimes involved a mixing of policy measures. For example, the 1995 revisions to Part L (the thermal part) of the UK building regulations included mandatory energy labelling in the hope that this would increase consumer awareness of energy efficiency and establish energy labelling for new buildings on a nationwide basis (Raman and Shove 2000).

The view that end-users' ignorance is at the root of the problem is shared by government and environmental pressure groups alike. For example, while the Association for the Conservation of Energy consistently argues for greater government intervention to correct market imperfections, it also points to the problem that householders have little or no idea of the environmental costs of their energy use, or of the financial benefits of investing in energy efficiency. Friends of the Earth criticize the government for preaching to the already relatively knowledgeable, but join the chorus of those identifying lack of information, lack of available technologies and lack of financial encouragement as barriers to energy efficiency. For Greenpeace, as for other non-governmental organizations, the problem is not the techno-economic view of technology transfer, but rather the scale and scope

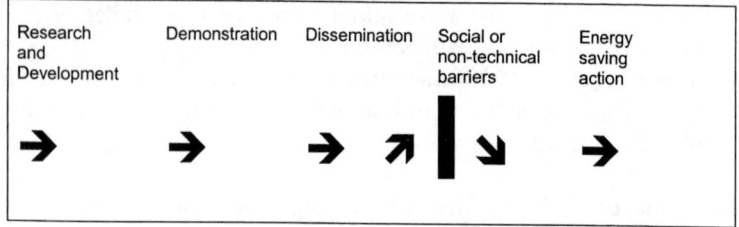

Figure 6 A techno-economic model of technology transfer

of the government's commitment to overcoming non-technical barriers. As these examples suggest, critics do not stray from the techno-economic view, but instead call for additional measures, often arguing for more substantial market intervention.

To summarize, technology transfer is popularly visualized as a struggle in which energy-saving technologies fight their way from the protected realm of research and development through the stages of demonstration and dissemination and on towards the real world of energy-saving action. Their path is strewn with obstacles and, as we have seen, any number of non-technical or social barriers threaten to prevent their successful application. Figure 6 represents this linear narrative in graphic form.

This is such a familiar perspective that it is only rarely articulated. However, Stephen Trudgill has formalized it in the form of a chain-linked model in which knowledge, technology, economic, social and political factors sequentially inhibit environmental action (Trudgill 1990: 4). As well as identifying these factors as sources of trouble, Trudgill suggests that solutions lie with the individual. As he explains:

> Motivation for tackling a problem comes from our moral obligation and our self-interest in enhancing the resource base and its life – thus enhancing, rather than destroying, planetary ecosystems and plant and animal species, including ourselves.
>
> (Trudgill 1990: 105)

This is a curious but nonetheless popular analysis of change and resistance. Although obstacles are often described and defined in rather generic terms, having to do with institutional features, such as the differing interests of landlord and tenant, financial incentives and the like, ways forward are generally thought to lie in the hands of key decision-makers and other autonomous individuals. The vocabulary is typically individualized. Phrases such as 'inadequacy of knowledge', 'technological complacency', 'economic denial or complacency', 'social morality/resistance/leadership', 'political cynicism/ideology', consistently feature in such debates. Levels of energy consumption in the built environment are believed to be the consequence of 'thousands' of individual judgements by 'property-owners and other decision-makers' (Olsson 1988: 17). All these are taken to be free agents, able to commit themselves at will to a more or less sustainable urban future. As Nigel Howard argues;

> ... there are lots of decision-takers. There are lots of people who have to be influenced, right from government to Local Authorities, developers, designers, material producers, professional and trade bodies. They all have a role to play. And what we have to do is try and influence all of them.
>
> (Howard 1994: 14)

This way of seeing energy and technical change positions more or less rational individuals as both the answer to, and to some extent the cause of, energy-related environmental problems.

In conclusion, the techno-economic model provides an organizational logic or operating principle upon which much energy research and policy proceeds. It represents a unifying force, linking the paradigms of policy with those of technical research, and equipping both with a quite specific view of choice, change, and the role of individual decision-makers. It functions as a self-sustaining, mutually reinforcing package of beliefs. Each element of this pervasive bundle – the transferability of technical knowledge, the individualistic theory of technical change, the sequential logic of research and development, and the implicit distinction between the social and the technical – feeds into the next, creating a web of belief strong enough to encapsulate technical researchers and policy makers, and elastic enough to span countries and continents.

Leaping the barriers

Widespread acceptance of the techno-economic model of technology transfer has marginalized other more structural explanations of innovation and change. As Thomas Hillmo observes, the notion of barriers suggests that 'problems are absolute' and that 'some actors' (i.e. scientists) 'know the truth about a problem', while 'other actors' (i.e. non-scientists) 'do not and obstruct the solutions in different ways' (1990: 124).

Although the role of the social scientist has become increasingly important, it is still typically limited to that of evaluating promotional techniques and undertaking surveys of people's attitudes in order to inform advertising campaigns and other such initiatives. This work has concentrated on what are referred to as the behavioural or human dimensions of energy efficiency (Stern and Aronson 1984; Rosa *et al.* 1988). Although drawing upon intellectual resources from economic sociology and environmental psychology, human dimensions research helps sustain the techno-economic model of energy efficiency. It too assumes that energy-saving actions are the consequence of informed rational action on the part of individual decision-makers. While differentiating between types of energy consumers,[15] the narrow focus on individual attitudes and motivations prevents further analysis of the social organization of decision-making or the construction of choice.

As a result, research into the human dimensions of energy efficiency has had relatively little to say about the consumption practices of organizations, or the role of institutions and groups involved in designing, financing and developing buildings. As Janda points out, human dimensions research 'usefully delineates differences in individuals' attitudes, (but) . . . does not reveal where these differences originate, how they develop, or if they can be changed' (Janda 1998: 31). Studies that abstract and analyse the opinions and motivations of designers and consumers isolate and freeze what are always contingent practices. As we go on to argue, this behavioural view of technical change fails to recognize the routine complexities of energy-related decision-making. In particular, there is little room in this model for an explicit recognition of the inter-relationship between technical innovation and social process, no space in which to consider the structuring of options and the intersection of competing priorities, and no way of recognizing the extent to which energy-efficient choices are embedded in the routines of domestic life and commercial practice (Shove *et al.* 1998; Shove and Wilhite 1999).

Changing course

We have now reached a turning point both in this book and in the history of energy research. The story to date has been one of successive refinement of what we have referred to as the techno-economic paradigm. We have shown how energy research programmes have been built up and how they have collectively reinforced a particular theory of choice and change. The institutions of research funding, together with the conventions of scientific practice, have hooked up with a model of energy policy and together focused on the production of transferable knowledge and the analysis of non-technical barriers to its effective implementation.

Having described some of the theories that lie behind conventional approaches to research and policy, we are in a position to take another tack. In the chapters that follow we highlight the limitations of the techno-economic model and illustrate the value of a more sociologically informed approach to the analysis of energy-related building design. We do this with the help of three case studies, each of which re-visits a specific aspect of the techno-economic paradigm.

Before embarking on these discussions we assemble a toolkit of relevant ideas and concepts drawn from sociological literature in the field of science and technology studies. Our rather pragmatic aim is to appropriate and, where necessary, adapt such ideas in order to develop a more robust analysis of energy efficiency in the built environment.[16]

A first move is to recognize science as a socio-cultural phenomenon, not as something that exists ready-formed outside society. Such a move has any number of implications. Amongst others it brings into question the 'authority of science', locates 'knowledge-claims in their social context' and identifies the 'relationship between such contexts and wider economic and political processes' (Webster 1991: 14). Having acknowledged the embedding of science in society, numerous scholars have gone on to ask 'What accounts for the emergence of particular technologies? Why do they appear when they do? What sorts of forces generate them? How is the choice of technology exercised?' (Volti 1992: 35). Traditional explanations look inwards, focusing on developments within the technology itself or following the ideas of famous scientists, inventors and entrepreneurs. Others look outwards, searching for explanations through further understanding of how societies influence the types of technologies they produce and asking, quite specifically:

> What role does society play in how the refrigerator got its hum, in why the light bulb is the way it is, in why nuclear missiles are designed the way they are?
>
> (McKenzie and Wajcman 1985: 2)

Challenging the notion that technological change has a logic of its own, these authors investigate the social contexts within which design solutions emerge and patterns of consumption evolve. Extending this point, Hughes argues that sociological, techno-scientific and economic analyses are permanently woven together (1988). As his analysis of the development of electricity in the USA graphically illustrates, the sometimes conflicting interests of social groups and institutions have to be considered alongside technological artefacts, both being implicated in the active fashioning of the future. Rather than understanding change in terms of individual belief, apathy or enthusiasm, it is instead necessary to understand the ways in which the social organization of energy-related choices structures opportunities for energy-saving action. Rather than assuming an unstoppable course of technological development, social (and technical) actors are taken to have a creative part to play in the construction of socio-technical trajectories. This is an important step because it positions energy research on what Michel Callon calls a new terrain, that of 'society in the making' (Callon 1987). As Webster has observed, this is a 'contested terrain, an arena where differences of opinion and division appear' (1991: 1) and one that must be analysed as such.

What does all this mean for the science and policy of energy efficiency? We take a strong line here. If we are to go along with the ideas sketched above we have to reject all forms of technological determinism: it is no longer possible to view technical innovation as a handmaiden to an energy-efficient economy. Second, we should recognize that the value of distinguishing between technical, social, economic and political aspects of energy use is limited, even misleading because each intersect in what Hughes has described as a 'seamless web' (Hughes 1988). Finally, we should abandon individualist explanations of technical change, including those that rely upon rational consumer action.

Adding these ideas together, the implication is that energy research should take note of those social and technical networks that frame the meaning of energy and that create more and less amenable landscapes of opportunity for the promotion of energy efficiency. Such landscapes are likely to vary over time and within specific settings and situations. Furthermore, opportunities for efficiency and conservation do not simply arise and recede according to some technological or economic logic. They are actively constituted by a range of non-energy considerations including design conventions, forms of investment analysis, theories of space utilization, concepts of the normal 'home', and so on. In other words, designers, technologists and energy researchers are together involved in constructing one another's choices. This idea hints at the malleability of socio-technical systems and reminds us of the networks of power at play, even with respect to the built environment.

Technical researchers frequently view buildings in terms of their physical attributes. As such they feature as 'essentially static objects' formed in a relatively standardized manner from an assembly of inter-connecting construction materials (Groak 1992: 6). Seen in this way, buildings give an impression of technical homogeneity: aspects of form and performance can be defined and described irrespective of geographical location, patterns of ownership, occupation or operation. But what if we view buildings as the material products of competing social practices? Again this suggests the need for a new way of thinking. For social scientists, such as Bruno Latour, understanding 'what machines are' is the same task as understanding 'who the people are' that shape their use (Latour 1987: 259). From this perspective, technologies and technological practices are 'built in a process of social construction and negotiation', a process driven by the social, political and commercial interests of actors implicated in the design and use of technological artefacts (Bijker and Law 1992: 13). It is true that buildings seem to be more complex artefacts than such discrete objects as the bicycle, the fluorescent light, or the refrigerator, all of which have been examined by scholars working within the field of science and technology studies. But there is nothing to stop us adopting a similar analytical stance and also tracing the characteristics of the 'actor world' that 'shapes and supports' the production of more and less energy-efficient buildings (Bijker *et al.* 1987: 12). Adopting this perspective means relating the form, design and specification of more and less energy-efficient buildings to the social processes that underpin their development.[17]

This emphasis on the social organization of design and use is at odds with the techno-economic perspective, which assumes that a 'repertory of well tried technical solutions' provides 'reliable precedents for designers' (Groak 1992: 6). Designers' strategies differ according to climate and culture, and their solutions are adapted to suit different circumstances. However, all appear to confront the common problem of accommodating the seemingly universal need for shelter and comfort. This analysis positions designers in a rather passive role, simply doing their best to meet societal needs. Such a perspective obscures other compelling questions, for instance, how are those needs defined, what constitutes comfort and what part do building designers play in constructing the needs they seek to meet? Similarly, questions about the design routes that are and are not taken touch upon more contested and controversial issues regarding the vested interests at stake in the shaping of the built environment.

Pursuing this theme, we might ask how rival schemes are valued by different participants in the design process and we might usefully consider how relations of power and processes of negotiation and compromise frame the end result. Rather than supporting a linear model of innovation, sociological research of this kind promises to reveal the web of interests at play, also identifying the paths not taken and the multiplicity of routes available at each twist and turn of the innovation journey.

This point draws attention to the neutral idealism of the techno-economic model, a model in which informed and rational individuals are expected to make use of a growing stock of technical knowledge. The techno-economic model does allow for differences of power at least to the extent that individuals are known to occupy positions of more and less influence. Promotional campaigns consequently focus on key decision-makers; designers and top managers; technicians and lower management; or domestic consumers. Each is treated differently and each is assumed to have a different part to play in a relatively static social hierarchy. This focus on the individual makes it difficult to appreciate the relative nature of power, or to take note of the fact that capacities for action differ from one spatial and temporal situation to another.

The paragraphs above introduce a number of critical ideas: promotional campaigns position technical knowledge as an irredeemably social phenomenon; they challenge individualist explanations of innovation; they point to the 'seamless' web of socio-technical change; they remind us that there are multiple possible pathways of innovation; they identify contested and competing interests; and they suggest that innovation is an active social process marked by situationally specific patterns of power. More prosaically, they suggest that technical choices are not solely determined by the knowledge or motivation of those involved. Two insights are especially relevant here. One is that knowledgeable designers and consumers are likely to have competing or conflicting priorities that complicate their energy-related decisions. The second is that choices and options are themselves socially structured.

With reference to the first point, Rick Wilk emphasizes the tremendous plasticity and flexibility of human action, arguing that it is always guided by multiple motives. Illustrating this point in terms of domestic decision-making, he reviews some of the overlapping logics at play in making what appear to be energy-related choices.

At particular stages in life, when the last child has moved out to go to college, for example, people were much more likely to repair or modify their houses. Couples often did not agree about temperature settings on thermostats, and ... discussions about home improvements involved a good deal of family politics. Sometimes changing the house had a lot more to do with making the marriage work than it did with saving money or energy. And we also found a significant number of people for whom energy saving was a political issue. Hatred of the nuclear industry led some people to cut down their electrical use; others hated being dependent on a huge corporation for their energy.'

(Wilk 1996: 149)

In the home as in the office, choices about energy are rarely what they seem.

Elaborating on the second point, that is on the construction of choice itself, Ruth Schwartz Cowan reviews the history of home heating and cooking systems in America. Following the introduction of cast-iron stoves, which replaced open hearth fires, she traces interconnections between stove producers, fuels suppliers and merchants, and maps their networks of influence through and between the worlds of production, wholesaling, retailing and the household itself. This exercise reveals the extent to which the householder, as decision-maker, is positioned within a framework of choice and impossibility. Such an analysis highlights the significance of the 'place and time at which the consumer makes choices between competing technologies', allowing Cowan to unpack the determinant of choices and observe the simultaneous emergence of technical pathways that 'seemed wise to pursue' or that appeared 'too dangerous to contemplate' (Cowan 1987: 263). This and other historical analyses have the further benefit of illustrating the rate at which contexts and circumstances change. As Cowan points out, 'Today's 'mistake' may have been yesterday's 'rational' choice' (Cowan 1987: 261).

Again we pause to review what this means for the analysis of energy efficiency. Wilk's observations highlight the risk of focusing on energy alone and of taking decisions at face value. In the context of building design, this suggests that we should not isolate questions of energy efficiency but should consider the ways in which such issues touch upon and reflect apparently un-related priorities. Cowan's analysis also prompts us to widen the scope of enquiry. As well as thinking about end-users or designers, it is clearly necessary to take account of the ways in which energy-related choices are embedded in the manufacture, distribution and retailing of technologies and in the commercial processes framing building design and development.

Like Bijker on fluorescent lighting (1995), and Cooper on air-conditioning (1998), Cowan examines the active construction of options and the manufacturing of demand. All three of these authors consider the process of diffusion as a process of invention and all three acknowledge the effort invested in creating and constructing 'need', whether that be for incandescent lighting, for central or stand-alone air conditioning systems, or for new forms of heating and cooking. Rather than assuming the adoption of such technologies and then seeking to understand the obstacles that stand in the way, the creation of demand is itself part of the story. This is often a contested and uncertain business. Cooper's history of air-conditioning in America and Wilhite's discussion of Japan and Norway both suggest that consumers had to be won round to the idea that it was good to have a centrally controlled, mechanically ventilated environment (Wilhite *et al.* 1996; Cooper 1998). The linear model of technical change, in which research flows into practice unless impeded by some non-technical barrier, fails to capture the social negotiation of need or the structuring of choice represented in these more sociological analyses.

In moving away from the linear model of technical change, these authors implicitly problematize the concept of barriers (Shove 1998b). Although we can observe more and less successful technologies, we no longer need to explain failure in such negative terms. It is not that new ideas are prevented from realization by a thicket of non-technical obstacles. Instead, the simple suggestion is that such ideas have yet to become enmeshed in the necessary socio-technical networks, that demand has yet to be created, and that for all sorts of explicable social and cultural reasons their adoption simply does not make sense in the specific context in question. As we argue in Chapter 8, the notion of treating existing social and cultural conditions as 'obstacles' is not especially positive or helpful, either for energy research or for energy policy.

For now it is enough to show that in borrowing ideas from the sociology of science and technology, we will also be borrowing a significantly different theoretical and methodological framework. The signs of this shift are already evident in the terminology we have begun to adopt. The talk is not of key decision-makers but of relevant social groups; not of technological diffusion but of the dynamics of socio-technical change; not of rational choice but of the construction of demand; the conventions of normal practice replace non-technical barriers, and so on. This represents much more than a change of semantic clothing. As will become apparent, the ideas hinted at above tap into theories of choice and change, which are quite unlike those embedded in the techno-economic paradigm that has hitherto dominated the field of energy research.

There are important differences at the level of social theory. But in borrowing concepts from science and technology studies and putting them to work with respect to the energy efficiency of the built environment, we are dealing in practicalities as well as theoretical abstractions. As we noted above, energy research is at a turning point. Policy makers and others are increasingly concerned that the techno-economic paradigm is simply not delivering either the research or the action required to transform patterns of energy consumption on any significant scale. Confounding barriers and obstacles appear at every turn. The search is on, now more than ever before, for new ways of thinking about energy consumption that might complement, or perhaps replace, the conventional techno-economic framework (Shove *et al.* 1998).

Alternative theoretical paradigms and models also imply new agendas and new divisions of labour. Research into the so-called 'human dimensions' of energy use typically concentrates on the attitudes and beliefs of end-users and consumers. Framed by a linear model of technological change, such 'end-of-pipe' social science promises to offer market intelligence, perhaps also identifying non-technical barriers and obstacles to the successful uptake of proven methods and measures. In the next three chapters we carve out, explore, and articulate a very different role for social research: one that investigates the structuring of demand, shows how specific configurations of building design and use favour some but not other forms of energy-saving action, and promises to identify socially viable pathways of change.

Instead of focusing on individual decision-makers we emphasize the contexts in which choices and options are defined and made. Questions about the construction of demand are critical. Instead of itemizing the barriers to technical change, we consider the localized circumstances in which energy-saving actions do and do not make sense. We highlight the social dynamics of innovation and reject neatly linear models of technological change. In doing so we make use of concepts that have been developed and refined with respect to various debates in the sociology and philosophy of science and that have generated and drawn from an extensive repertoire of empirical research. In these respects our approach is not especially novel.

The cases we take and the examples we give are selective and partial; but in bringing ideas from science and technology studies to bear upon three key areas of energy-related action we show how sociological analysis can reinvigorate both the theory and practice of energy and environmental policy.

Following energy efficiency

The next three chapters are located in the worlds of building design and technological development. Having left the realm of building science and technical research, we follow issues of energy efficiency as they appear in practice and as they emerge in different sectors and organizational settings. Questions of energy efficiency rarely crop up in 'pure' form in the real life situations we now examine. In methodological terms, we have chosen to follow and describe the patterning of priorities and the non-energy-related considerations that make a difference to energy performance. This takes us into new territory, for instance, leading us into discussion of the politics of nuclear power; of the dynamics of the UK's commercial property market; the ambitions of one-man-band insulation installers; and the history of public sector investment in housing.

The goal of understanding energy conservation in practice could be pursued in any number of ways. Our approach reflects two further ambitions. One was to undertake a programme of empirical research that engaged with core elements of the conventional techno-economic paradigm. The expectation that proven cost-effective energy-saving technologies will diffuse unless impeded by some non-technical barrier represents a central theme. What could we learn about the practicalities of technological diffusion by tracking the development and adoption of one very simple energy-saving technology? That was one question. The second related to the notion of information deficit. Do adequately informed decision-makers put their energy-saving expertise into practice and is lack of energy-saving action a symptom of ignorance? Investigation of those involved in housing design in the public and private sectors promised to address this question. Finally, what could we learn about the idea and the operation of non-technical barriers. Focusing on one often cited barrier to energy-saving action, that is the gap between landlord and tenant interests (the former want to limit capital expenditure and it is the tenant, not the landlord who benefits from the energy savings that follow), we sought to understand the dynamics of this relationship and its implications for energy-related investment.

Our second ambition was to develop a more sociological framework for the analysis and understanding of energy-related practice. In this we make use of many of the ideas outlined above, but often in a rather understated way. Although the accounts we offer are informed by these perspectives the next three chapters are not explicitly designed to advance the field of science and technology studies. We reflect on the implications of our approach in the final chapter, but until then we concentrate on the narratives and accounts of energy in action as provided by people involved in the insulation industry, in housing and in commercial office development.

As we saw in Chapter 3, it is not always easy to determine exactly what constitutes an energy-saving measure or to know what its benefits might be. Passive solar design is a case in point. Fortunately, some energy-saving 'technologies' have much more readily identifiable environmental benefits. Insulation is one. Although available in different forms (for instance, mineral fibre, foam, polystyrene, etc.), insulation represents a stable, proven and almost always cost-effective measure. On this basis, one might expect to see a steady pattern of diffusion across the European building stock. In fact, levels of insulation are extremely uneven. Taking this observation as its point of departure, Chapter 5 reviews differing explanations of why this might be so. Techno-economic accounts focus upon differences of climate, cost and political commitment, but take little note of the part that the insulation industry itself plays in creating and constructing demand. Chapter 5 begins by charting histories of insulation in France, Denmark and Sweden, relating these to the structure of the industry and the extent to which manufacturers interact with political and consumer interests. This comparative review is necessarily static. In order to show how the roles of government, industry and consumer evolve over time, we offer a more detailed account of cavity wall insulation in the UK. The stops and starts of this story and the erratic narrative of international diffusion illustrate the limits of conventionally linear theories of technological change. In this case we see how the culturally and temporally specific interests of government, industry and consumer intersect, and how their localized configuration determines the fate of even such a simple and such a standardized 'technology' as insulation.

Having taken a closer look at competing theories of technical change through our analysis of insulation, we focus on the question of technical choice. Chapter 6 takes issue with the notion that decision-makers' actions reflect their knowledge and commitment to energy efficiency. Dissemination programmes and 'best practice' initiatives assume the transferability of the solutions they describe. In addition, the logic of economic rationality leads observers to expect properly informed decision-makers to put their knowledge into practice. There is an extensive reservoir of knowledge about how to produce more energy-efficient housing and although UK house designs differ, sometimes radically, expectations of what a house is and what it should provide limit the range of energy-related options and possibilities. Given a good stock of knowledge and a restricted range of choices this is as promising a context as any in which to observe the flow of science into practice. Taking housing as our point of reference, we followed the decisions and actions of individual (more and less knowledgeable) designers working for private sector housebuilders, Housing Associations and Local Authorities. Although all dealing with a broadly similar technical challenge – that is the production or renovation of dwellings – those we interviewed were situated within significantly different organizational environments. Priorities, perceptions of risk and value, and normal working practices varied accordingly. As a result, technical measures adopted in one context, for instance, in the world of Housing Association development, might be dismissed out of hand by a private sector builder. Technologies that were successful in one sector failed in another, even though the cost–benefit equations were similar, and even though all parties had the necessary knowledge and expertise. Although difficult to comprehend in terms of abstractly rational models of decision-making, these practices make sense when viewed from a more sociological perspective.

Having identified the ways in which organizational contexts frame opportunities for energy-efficient practice, we track the evolution of such contexts over space and time. Developing this theme, Chapter 7 challenges the typically static representation of market processes and market barriers. Taking the case of the landlord–tenant relationship as an example, this chapter shows how patterns of influence and power ebb and flow within the world of commercial real estate, and how these affect the details of building design and the form and specification of office development. We begin with a brief review of the changing structure of the British property market from the post-war period up until the boom of the 1980s, showing how this has influenced the development of standardized design specifications. Switching our attention to France we consider a different model of commercial property management and again show how this shapes 'normal' design standards and the process of technical change. Returning to the UK, we follow more recent movements in the property market, suggesting that these trends mirror localized tensions between investors, developers and occupiers within what is becoming an increasingly global context. Rather than talking about 'market barriers', we relate innovation in design and specification standards to the shifting structures of social and commercial influence. As we show, new configurations of power may inadvertently align the interests of development actors in ways that favour (or inhibit) the efficient use of resources in the built environment. By focusing upon the changing logic of environmental innovation, rather than on the knowledge and motivation of individual actors, we illustrate the explanatory weakness of techno-economic models, at the same time suggesting the relevance and power of the analytic tools we have borrowed from the sociology of science and technology.

Each case study has a distinctive focus, the first challenges notions of technological diffusion; the second re-visits concepts of choice and beliefs about the relationship between knowledge and action; and the third unpacks ideas about obstacles and barriers and offers a new perspective on the dynamics of energy-saving action. All three dramatize the diversity of contexts, cultures and situations in which the standardized results of building research are and are not adopted, adapted and appropriated. This has the double effect of highlighting tensions between global or cosmopolitan knowledge and the localized settings of action and practice, and of raising practical questions about more and less socially viable strategies for promoting energy conservation.

5 The politics of insulation

Technological fixes are unlikely to provide all the answers to recognized environmental problems, but there are some instances in which simple measures have positive environmental benefits. Improving home insulation represents one of the single most effective measures a householder can take and, in terms of new building construction, better insulation promises to pay for itself in terms of future energy savings (Department of the Environment 1991).[18] Despite this being a proven and well-established technology, existing levels of insulation in the UK and in other northern European countries do not correspond to those justified on techno-economic let alone environmental grounds. In simple terms, there is an 'efficiency gap' between current practice and that which is technically and economically sensible. Why is there this gap, how can we explain national variation across Europe and how and why are levels and rates of insulation as they are?

Conventional explanations for the differential use of insulation generally focus upon climatic conditions (by implication, colder countries should be better insulated because the economic benefit is likely to be higher) or political will (by implication, levels of insulation will be higher where governments overcome market barriers through regulation, or by means of grants, subsidies and information campaigns). Our research suggests that the picture is altogether more complicated. Local, historical, cultural and commercial factors have a bearing on standards of insulation and the rate at which this technology diffuses over time.

This chapter examines the history and practice of building insulation from two angles. The first part puts climatic/economic and political/market failure explanations to the test with reference to a comparative review of insulation in France, Sweden, Denmark and the UK. The second part homes in on the erratic history of cavity wall insulation in the UK. Although cavity wall insulation is of undoubted economic benefit, rates of filling have not increased as steadily as one might expect. It is difficult to represent the jagged peaks and troughs of insulating activity as a narrative of technological diffusion and again other more sophisticated explanations are required. In both cases we suggest that the patterns we observe are the result of inter-dependent socio-technical networks in which the industry itself has a critical part to play. In short, climatic conditions, political will, and technical and economic benefit are only part of the picture. More than that, their influence is modified and mediated by local, historically specific conditions and circumstances. The final section considers the implications of these comparative and historical analyses for the development and application of other standardized energy-saving measures.

Conventional explanations

Any really major reduction of carbon dioxide emissions will require an equally dramatic improvement in the energy efficiency of the building stock and in northern Europe this means more and better insulation. Some countries appear to have a head start. The levels of insulation expected in the UK are, for instance, much the same as those required in Sweden some thirty-five years ago. Such simple comparisons are inappropriate. Even so, a discussion of national differences may help to explain the process of technological diffusion. At first sight, climate has a lot to answer for. Friends of the Earth suggest that the building regulations of Scandinavian countries demand higher standards of insulation because they have colder climates (Friends of the Earth 1990). If climate was the sole determinant, the most stringent thermal regulations would be found in the coldest countries, yet the mineral wool industry's own research shows that this is not the case (EURIMA 1990, 1991). Although Scandinavian countries do come out on top, France has a higher ranking than either Austria or Switzerland, both of which figure around the middle of the league table. Although Denmark and the UK have very different standards of insulation, the industry's literature also emphasizes their climatic similarity (EURISOL 1981). Similar does not mean the same, and in building terms annual averagesare irrelevant if the temperature occasionally plunges to life-threatening levels. Nevertheless, climate does not have simple consequences for insulating practices.

Most obviously, climatic conditions have remained more or less constant across Europe. Colder countries have not suddenly got colder but they have become better insulated. The variable here is not climate alone, but climate together with changing concerns about the costs and sources of energy, which brings us to a second relevant consideration, namely that of political interest in energy supply and environmental change. Political explanations for the differential uptake of insulation suggest that this is a consequence of government intervention whether through regulation or subsidy. Thus the argument runs that Denmark has more insulation because insulation has been subsidized. Again there is a certain logic to the position but again it is misleading. Government programmes do not always have the desired result regardless of time and place, nor do they simply mirror national environmental commitments. Denmark has, for instance, introduced various grants, loans and tax rebates as part of its highly regarded conservation strategy. However, their impact seems to have depended less on the type and level of subsidy than on the timing and context of its implementation. As in Britain, subsidies introduced in the aftermath of 1970s oil crisis have been more successful than subsequent, equally attractive, initiatives launched in less traumatic times. Although building regulations can and often do have consequences for rates of insulation, it is again misleading to take these as a direct indicator of political will. Effective regulation is a delicate matter not least because it is so difficult to regulate ahead of current practice. For this reason alone, standards of insulation presume and reflect a measure of industry-wide consensus. Regulatory practices are influenced by technical research, and building standards are modified in the light of changing environmental commitments, but they are also enmeshed in local networks of commercial interest and conventional construction practice (Shove 1995; Wallace 1995; Shove and Raman 1996).

Our research, which involved interviews with insulation manufacturers, trade associations and building regulators in France, Sweden, Denmark and the UK, generated a much more complex picture of how insulating practices have developed and of the respective roles of government and industry. Although commercial interests rarely figure in conventional explanations of technological diffusion, they proved to be extremely significant, not alone, but in combination with other considerations. The following sections review the distinctive blend of commercial and political interests that constitute the 'cultures of energy conservation' in each country and that help explain the uneven distribution of insulation around Europe.

Cultures of energy conservation

Taken as a whole, Scandinavia is frequently referred to as an exemplar of effective energy conservation. On closer inspection, strategies within Scandinavia have different histories, aspects of which are also found in non-Scandinavian countries such as France. Discussion of three national environments shows how emerging markets for mineral fibre insulation have been shaped by a combination of natural factors (climate, the size of the country, the distribution of resources, etc.) mediated and modified by changing political contexts and government-industry relationships.

Denmark

Denmark's enviable position at the top of the insulation league table is generally explained in terms of its climate and its relatively high level of prosperity. But Denmark has not always been so green. Before the 1973 oil crisis, Danish housing came second only to the American stock in terms of thermal inefficiency (Danish Energy Agency 1992). What explains the sudden shift from a culture of consumption to one of conservation? The rate of change depends in part upon the existence of a remarkably small population of energy-related professionals (see the discussion of 'close communities' in Chapter 2). The knowledge and interests of industry, government and the research community overlap to a much greater extent than in larger, less homogenous, countries such as the UK.

The resulting patchworks of personal experience help sustain a shared sense of historical development. For example, respondents' stories about wartime shortages and bitter winters foster an image of government and industry working together in pursuit of a common goal. As described, these events led to the establishment of the Danish Building Research Institute (1948) and later the Thermal Insulation Laboratory (1959), also inspiring industry sponsored research and in turn leading to government recognition of the value and technical feasibility of insulation. Comfort, not cost, was the prime concern, but whatever the motive, key players slotted into place in the post-war years. The insulation industry was established with a proven product and an assured market thanks to the development of building codes devised in conjunction with a sympathetic state. The oil crisis took place against this background. Building on an already established network of expertise, the ambitious Danish energy plan of 1976 sought to reduce heat demand in new and existing building. Large-scale programmes of government subsidy for insulation and other energy-saving measures soon followed.

Commercial and political interests have not stayed still and other issues, for example, debate about the development of nuclear power and more recent environmental anxieties, have changed the identity and relative influence of the parties involved. Through the 1960s, the Danish utilities had been happy to rely upon cheap oil. The 1970s energy crisis led to a change of opinion initially in favour of nuclear power. But the changed political circumstances also produced a new grass-roots interest in the form of the Danish Organization Opposing Atomic Power (OOA). The OOA was broad-based, well-organized and widely supported. It developed alternative energy plans and by means of vigorous campaigning and technical debate, convinced the government to provide further support for energy conservation.

The insulation industry formed new alliances as the rationale for energy conservation swung from comfort to cost, and from fuel security to environmental protection. Each shift brought in a new population of institutional and organizational players, each marked by a different configuration of priorities. Although the range of players and priorities has evolved over time, events continue to be shaped by long established links established between academia, industry and government. Over the last 30 years the mineral wool insulation industry has collaborated with other actors in an effort to curb consumption and reduce Denmark's reliance on imported oil. As Nielson explains, 'this has meant that investors (supply companies, industries, single consumers etc.) have been able to act in confidence to secure a firm national strategy' (1990: 82).

As elsewhere, the worldwide downturn in construction has hit the Danish mineral fibre industry, resulting in redundancies and a streamlining of operations. But because of their recognized place as key actors in the formulation of energy policy, fibre manufacturers have had an active part in developing their own internal markets, in the process acquiring the experience and confidence needed to expand abroad. These social links have lent the Danish insulation industry a measure of institutional legitimacy.

Sweden

As in Denmark, Swedish energy experts belong to a close and familiar professional community. Informal contacts are again crucial as are shared experiences and common histories within and between industry, government and academia. If anything, state involvement in the insulation industry has been greater and more consistent than in Denmark. Benefiting directly from government sponsorship and subsidy, insulation manufacturers have also been able to rely upon long-term programmes of building construction.

There are, however, two important points of difference. First, Sweden's climatic conditions influence the organization of construction as well as the quality of its products. The typical pattern is one in which dwellings are constructed from standardized pre-fabricated components that are shipped to their final destination. Houses are then erected in a matter of days during the relatively short out-door construction season (Schipper *et al.* 1985). Designs must be robust enough, and insulation levels high enough, for the same product to function as well in Lapland as it does in the more amenable climates of southern Sweden. To some extent, high standards of insulation simply reflect this commercial–climatic requirement.

Second, and in greater contrast to Denmark, Sweden used to be committed to nuclear power generation. Widespread use of relatively expensive electric heating favoured the production of well-insulated homes and again generated a secure market for insulation well before the oil crisis. The Swedish government was also able to draw upon versatile and experienced manufacturers when developing energy plans and conservation strategies in the mid and late 1970s. The difference is that these energy-saving initiatives formed part of Sweden's nuclear energy policy and were not devised as an alternative to it, as in the Danish case. The result of Sweden's 1980 referendum on nuclear power changed the rationale for energy conservation, and programmes have since been designed to ease the transition to a non-nuclear future.

In short, the Swedish story revolves around greater or lesser commitment to nuclear power. Either scenario favours the insulation industry (albeit for different reasons) creating a situation in which mineral wool manufacturers have enjoyed regular and steady state encouragement. By contrast, the Danish industry has experienced a rather more unsettled past, interests in insulation swinging more markedly as the emphasis shifted from energy security to environmental concern.

The final comparison with France illustrates a further combination of interests. Reliance upon electricity and nuclear power are again significant factors, but differences in the management of energy supply have further consequences for the promotion of energy efficiency.

France

The French have placed their faith in nuclear power. Equally important, electricity supply is dominated by the state monopoly, the Electricité de France (EDF). French electricity is among the most expensive in Europe and the EDF has an immediate interest in energy conservation as a means of providing energy services (such as heating and cooling) at competitive rates. In this curious situation, the EDF is in a position to monitor, promote, and sometimes enforce the use of insulation. No wonder, then, that alliances between insulation manufacturers and the EDF were formed well before the oil crisis.

Despite climatic differences, despite a more traditional and a more localized approach to building construction and despite a more dispersed community of researchers, government officials and manufacturers, France has developed thermal regulations that approach the standards adopted in Denmark and Sweden. These depend upon a different but no less effective community of interests, this time between the EDF, the government, and the insulation industry. The resulting network of inter-dependencies provides the organizational backbone for subsequent energy plans (International Energy Agency 1991: 147). Situated within these interlocking interests, the French government has devised and managed a seemingly co-ordinated approach to energy efficiency that draws upon, develops, and exploits experience and expertise within the insulation industry.

As in Scandinavia, the French insulation market has been shaped by government regulation linked to a national energy plan and is in turn influenced by the interests of the insulation industry and those of the EDF. Rather than being driven by political concern for the environment, as in Denmark, or by the need to reduce demand as nuclear power is phased out, as in Sweden, sales of insulation in France are more directly related to the cost of electricity.

In all three environments, much depends upon the relationship between government and industry. In the small Scandinavian countries, this is fostered by a shared professional culture. In France, the EDF and the insulation industry are united by a common yet differently motivated interest in energy conservation. Either way, the external influence of climate is mediated through these cultures of conservation. In addition, the notion that standards of insulation simply reflect levels of environmental commitment is misleading. As we have shown, patterns of diffusion reflect the interlocking of political and commercial interests.

The European insulation industry

National markets have their own histories and characteristics. However, commercial opportunities transcend national borders and cut across environmental-political cultures. Insulation manufacturers, such as Rockwool, Saint-Gobain and Owens Corning, currently operate in more than one country and the history of their international development is itself revealing. Rockwool, started in Denmark, and Saint-Gobain from France, took a lead in early research and development and their exports dominated the pre-1973 market. As we have seen, the oil crisis changed national and international commercial landscapes beyond recognition, transforming actual and potential markets as governments and individual consumers rushed to reduce energy consumption. In response, Rockwool and Saint-Gobain opened new factories throughout Europe, competing with local manufacturers such as Pilkington in the UK. As the subsequent evolution of these companies shows, the balance of national and international interest in the production and consumption of insulation shifts continually and in ways that are not simply related to the environmental policies of individual nation states.

The European Union provides another important context for transnational industry-related influence. By acting with European regulators and with environmental non-governmental organizations the European Mineral Wool Trade Association (EURIMA) seeks to situate the industry within a wider community of environmental politics, a community that has the potential to exert some influence over the shape and size of national markets. Whether or not this influence is realized, the process of Europe-wide negotiation has the indirect effect of creating and maintaining an informal network linking national and international players within the industry.

We began this discussion with the observation that the actions and interests of insulation manufacturers were routinely overlooked in climatically or politically determinist explanations. The three examples we have examined suggest that there are significant national differences in the way in which insulation has been developed and adopted, but that the key factors are not the ones usually cited. We have argued, instead, that much depends upon the form and strength of government–industry interaction. This in turn relates to the detailed organization of building construction, the structure of the insulation industry both locally and internationally, the management of energy supply, and the role of different government agencies.

In the next section we review the history of cavity wall insulation in the UK in similar terms. How do government and industry interact and what does this mean for the rate at which homes are insulated? This more detailed discussion complements the comparative analysis developed above, introducing further questions about the relationship between government and industry and the role of individual consumers.

Filling the gap

Most houses built in Britain since the 1930s have cavity walls[19] (Brunskill and Clifton Taylor 1977) and cavity wall insulation, of whatever form, is a particularly effective energy-saving measure. The National Cavity Insulation Association claims that 'Full use of cavity wall insulation could cut power requirements for heating buildings by one-quarter – the output equivalent of two major power stations' (NCIA 1989: 1). Despite these advantages and despite the fact that cavity wall insulation promises to save householders up to a third on their heating bills, the Building Research Establishment's Energy Conservation Support Unit estimates that there are something like eight million homes in the UK that have unfilled cavities.

Explanations for this technically and economically unwarranted situation tend to focus upon the consumer. Given the obvious advantages of cavity fill, the assumption is that householders do not yet know of the benefits, or that they are harbouring unfounded fears about insulation or the companies that install it. In response, the UK government has sought to hasten the uptake of cavity fill by providing expert advice and information and by undertaking research to assess technical risk and demonstrate economic benefit. These strategies, designed to push the UK up the diffusion curve of technological progress, do not take account of the wildly erratic history of the cavity wall insulation industry, nor of the changing preoccupations of potential consumers.

It is impossible to recover the data required to produce a precise graph of insulating activity, but there are undisputed trends in the rate at which cavities have been insulated from 1960 to the present day. These are illustrated in Figure 7 above.

Why have sales of such an energy efficient and such an environmentally important commodity had so uneven a history? Standard models of technological diffusion cannot account for the timing and the scale of these ups and downs in the industry's history, nor for the slowing of current activity. If we are to make sense of this history we must again attend to the ever changing relationship between

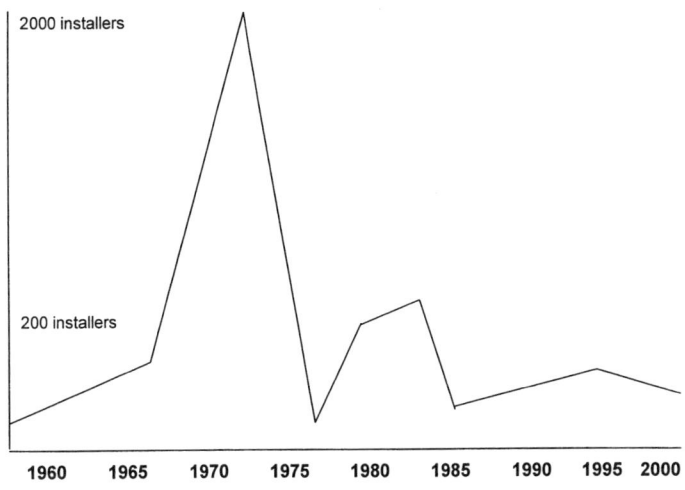

Figure 7 Activity in the UK cavity wall insulation: 1960–2000

consumers, industry and government. Analysis of relevant trade journals and interviews with twenty-four insulation installers, five manufacturers and all four UK cavity wall insulation trade associations allows us to follow the industry's uneven history, looking at the distinctive combinations of commercial, regulatory and market pressures reflected in the peaks and troughs of the graph.

First fillings 1959–74

Once built, it is difficult to reach the space between the inner and outer layers of a cavity wall. The range of insulating materials that can be pumped, injected or blown into this gap is limited, but one of the first to be tried in this role was urea formaldehyde foam. The earliest foam entrepreneurs had to overcome technical suspicions about a product that was invisible once installed and of marginal economic value given the relatively low cost of fuel during the late 1950s. For the first 15 years of its career, the cavity wall insulation industry consisted of a few major chemical companies (British Industrial Plastics, Ciba Geigy and Bordon Chemicals) selling raw materials to small-scale enterprises prepared to try their hand at selling and injecting foam.

The sudden rise in heating costs combined with a real fear about long-term energy supply had a dramatic effect on the market. In

1974, the heyday of the insulation industry, the National Cavity Insulation Association had 2,000 applications for membership. At that time it was possible to set up in business with no more than a ladder, a van, an 'installation kit costing £1' (Leeden 1975: 1166) and a franchise from a chemical supplier. In the words of one installer, 'all you had to do was walk down the street and ring a bell and everyone would come out.' The sheer scale of door-to-door selling guaranteed success and this brief 'cowboy' era undoubtedly did a lot for energy conservation. The more sales persons there were the better the insulation sold, and the more profit there was to be made the faster the rate at which people moved into the business.

This upward spiral of activity had other consequences. It is difficult to think of a better context for sharp practice and downright cheating for householders have no way of checking that cavities have been properly insulated. Unscrupulous installers over-diluted the foam mixture, failed to drill enough injection holes, or drilled all the holes but then never installed the foam. Even honest salesmen cashed in on consumers' fears about the size of future fuel bills and adjusted their rates accordingly. One way or another it was easier to make a quick profit in insulation than in other more regulated industries. For a time, installing cavity wall insulation was an attractive prospect for a rather mobile population of small-scale entrepreneurs. These people made a very effective job of promoting insulation despite, and perhaps because of, their 'cowboy' tendencies.

Householders bought cavity wall insulation at an unprecedented rate in the mid 1970s because there were so many people selling it to them. To understand why this is no longer the case we need to consider subsequent developments within the industry.

Curbing the cowboys 1975–81

Chemical suppliers, who had benefited from the sudden increase in activity in the early 1970s, began to worry about their inability to control those who were installing their product. Fearing that public confidence might collapse, and with it the potential for longer term profit, the main suppliers invited the British Board of Agrément to monitor installers on their behalf. As it happened, control also came from another quarter. Thousands of people were making radical and potentially risky alterations to their properties[20] and these technical concerns led local and national government to reconsider their position. In 1972, cavity wall insulation was deemed to be a structural alteration and installers had to apply for type relaxation of the

Building Regulations before starting work. To begin with, seeking permission was a mere formality but by late 1974 many Local Authorities were exerting their right to refuse. Local Authority building control officers could prevent home owners from insulating cavities if they felt there was an 'unacceptably high' risk of rain penetration, but there were no strict criteria for establishing this risk and interpretations varied widely. The then Department of the Environment was put in an extremely awkward position. In refusing permission, Local Authorities prevented householders from improving the efficiency of their homes at just the moment when the Department of Energy launched its 'Save It' campaign promoting insulation and encouraging energy conservation.

The only thing to do was to amend the national regulations, but by that time many installers had already left the field in search of easier ways of making a living. Under the new rules installers had to give Local Authorities seven days notice but, providing this condition was met, providing the Authority had no objections, and providing that the installer and the insulation system had British Board of Agrément certificates, work could proceed. From this moment on it was virtually impossible for an installer to get work without a British Board of Agrément certificate. As a contributor to *Building* noted: 'for installers, business now hangs on that precious Agrément approval and it is proving very difficult to enter into the field' (*Building*, 5 March 1976: 101). By 1976 it looked as if the industry had indeed settled down and the steadily increasing number of successfully insulated houses was itself evidence of the reliability of urea formaldehyde foam and the value of cavity wall insulation.

Looking back, these first phases of the industry's history were shaped by different and often conflicting interests. From the chemical companies' perspective, insulation represented a new market for products already sold for other uses in other contexts. For their part, opportunistic entrepreneurs saw cavity fill as an easy route to self-employed success. Having no long-term commitment to insulation, installers had no reason to stay when things got more difficult. Local Authorities' selective refusal to grant permission limited the market and delayed response on the part of the Department of Environment inadvertently curbed energy-saving activity just as the Department of Energy sought to encourage it.

Fuel prices were still relatively high in 1975 but demand for insulation was not sufficient to entice really large numbers of installers back into the business once the type relaxation issue was resolved. In any case the more established companies and their suppliers, aided by government regulation, and British Board of Agrément certification, had by then closed ranks against the cowboys. The days of 2,000 plus installers were over for these moves secured what seemed to be a more certain future for around 200 increasingly respectable installing companies.

Foam fears 1981–83

This contentment was not to last. In 1981, Canada and America banned cavity foam insulation on the grounds that it had potentially harmful effects ranging from 'discomfort to the definite possibility of cancer' (*Building Design* 1 May 1981: 1). Over a million householders had already filled their walls with urea formaldehyde foam by the time the scare broke in Britain. Cavity fill installers still recall the day when order books emptied as potential customers rushed to cancel following national media coverage of the formaldehyde story by News-night, News at Ten, and Esther Rantzen. Reports of the time empha-size the unnatural capacities of this synthetic substance, fostering a nightmare image of a potentially carcinogenic monster creeping into the very fabric of the nation's homes. In Sheffield, urea formaldehyde foam 'burst through the plasterboard inner walls of the homes, mainly into cupboards and around skirtings' (*Building Design* 15 May 1981: 1) and in Hemel Hempstead, 'U-foam had forced its way into kitchen cupboards, pushing pans onto the floor' (*Building Design* 30 April 1982: 1). Cancer was the main concern but American research had also shown that formaldehyde fumes could cause immediate respira-tory problems. These could be very unpleasant. The Shearing family, for instance, 'spent their first insulated winter sleeping with every window open for ventilation' (*Building Design* 29 February 1980: 1). The environmental hazards of insulation surely outweighed the bene-fits in this case and who knows what the family's fuel bill came to that year.

There had been few recorded problems prior to the early 1980s (and foam had, after all, been installed in more than a million homes), but more and more cases came to light as the story devel-oped. It was certainly odd that no one had noticed urea formaldehyde foam could have such effects before and, if anything, this gnawing uncertainty exacerbated the situation (Cockram and Arnold 1984). Perhaps 'lack of recognition of formaldehyde irritation had concealed the extent of the problem' (*Building Design* 23 January 1981: 1) and perhaps there were unknown numbers of unknowing sufferers.

Well-founded or not, fear about the possible effects of urea formaldehyde foam had a profound effect on the industry. By 1983, only one major national cavity wall insulation contractor remained in the UK (*Building* 9 September 1983: 1). After the foam scare, out of work insulation contractors looked around for alternative materials and fibre producers such as Rockwool, Pilkington, and Gyproc, and polystyrene bead suppliers including Shell and BP, began to take over an industry initially dominated by British Industrial Plastics and Ciba Geigy. The demise of foam inspired mineral fibre manufacturers to develop faster and more efficient methods of filling cavities with chopped and treated edge trim generated in the process of making prime quality loft insulation. Subsequent competition between the main fibre companies, Rockwool, Pilkington (now taken over by Owens Corning) and Gyproc has had far reaching consequences for installing contractors and their customers.

Fibre wars 1984–92

The 1973 oil price rises inspired government action on a number of fronts. The homes insulation scheme introduced grants for loft insulation in 1979 and successive changes in the UK building regulations have increased demand for insulation. Sited in areas of high unemployment and benefiting from a range of government subsidies, mineral fibre plants were built at Pencoed (Rockwool, 1980), at Runcorn (Gyproc), also in 1980, and at Washington and Stirling (Cape). These factories produced vast quantities of loft insulation, the waste edge trim of which was used for cavity fill. Fibre sales overtook those of polystyrene bead in around 1984–5 and the recent history of cavity wall insulation is, in effect, a history of relations between the fibre producers.

While installers compete for sales, manufacturers compete for installers and, in particular, for those who sell a lot of material. Manufacturers are keen to help installers secure major Local Authority contracts, but the strategy of offering special terms to allow an installer to win work has its dangers. Such action simply encourages other manufacturers do the same. In 1991, manufacturers, desperate for market share, together with installers, desperate for work, were competing for Local Authority contracts at 'silly prices' sometimes as low as £70 per house. These price wars put enormous pressure on the installers' profit margins and as one explained, 'we could go bankrupt for a burst tyre.'

The cost of fibre is less significant for those working in the private domestic sector and, in any case, it is easier to adjust prices here than in the publicly competitive environment of Local Authority tendering. Skilled installers have been known to persuade householders to pay up to £800 for their services (more than ten times some Local Authority rates), making significant profits on the basis of relatively few installations. Such companies spend a greater proportion of their time selling as opposed to installing insulation and buy relatively small quantities of material at relatively high prices.

Bit by bit the gap between the public and private sector markets widened. Installers who sell to individual householders are less dependent on insulation manufacturers and, for their part, major fibre producers have no real interest in stimulating this market. The full weight of manufacturer support (and pressure) lands on larger installers exclusively devoted to Local Authority contract work. Fibre price wars are fought in this context and, as a result of successive rounds of undercutting, Local Authorities are able to fill cavities for less now than they would have paid ten and even fifteen years ago. Installers choosing to remain in the domestic sector make little difference to the well-being of an industry that has become heavily dependent on the swings and roundabouts of Local Authority spending. The present rather bleak picture is therefore one in which domestic insulation is caught in a downward spiral. It is difficult to foresee a time when small-scale entrepreneurs will be attracted back in sufficient numbers to really make a difference to the individual consumer, yet that is what is required if those 8 million householders are to be persuaded to fill their cavities.

Government, industry and consumer interaction

The figure below summarizes the involvement of consumers, of the government, and of installing and manufacturing companies over the last 40 years, italicized entries representing what seem to be the key feature of each period.

This account has shown how commercial, government and consumer interests have combined to influence the fitful path of technological diffusion. What began as an offshoot of the chemical industry is now dominated by the economics of mineral fibre. The number of installing companies has changed by a factor of ten, cavity fill has been beset by technical scares and panics, the focus has shifted from the private to the public sector, and from door-to-door selling to Local Authority tendering. Putting the pieces together we get an extremely complex picture of differing and changing motives and concerns.

The story of cavity fill in the UK and histories of insulation in

	Installer	Government	Consumer	Manufacturer
Filling the Gap 1959-1974	*2000 companies, the 'cowboy' era*	Not involved	Highest levels of interest ever	Chemical companies
Curbing the Cowboys 1975-1980	200 increasingly respectable installers	*Type relaxation issue, concern to regulate the industry*	Interest drops with the drop in sales effort	Chemical companies
Foam Fears 1981-1983	Total collapse	Scientific uncertainty	*Health scare and consumer panic*	Polystyrene bead and mineral fibre
Fibre Wars 1984-	Division between domestic and contract installers	Indirect influence on Local Authority spending	Domestic market virtually abandoned by the industry	*Fibre price wars*

Figure 8 Changing interests in UK cavity wall insulation

France, Denmark and Sweden all suggest that energy-efficient action takes place *within*, not outside, localized, culturally and temporally specific settings. The idea that levels and rates of insulation depend upon the actions and interactions of a range of inter-dependent players has important implications for the conceptualization of technological diffusion. If we accept this analysis we should also abandon the notion that seemingly backward nations might adopt the practices of their more advanced neighbours. For sure, Swedish regulations require more insulation than demanded by current UK standards, but there is no reason to suppose that the mixture of relevant interests in the UK will ever correspond to that which underpins current practice in Sweden. It would be just as difficult to turn the clock back and somehow re-create the conditions of the early 1970s, conditions in which British householders were virtually queuing up to have their cavities filled.

Although appealing and although extremely influential, the vision of linear progress toward a better insulated and so more sustainable future simply does not square with the realities and practicalities of the insulation industries we have examined. We have instead described separate, relatively self-contained, scenarios each driven by its own history and logic and each revolving around a distinctive combination of consumer–government–industry interests. Rather than a neatly ordered progression of sequential steps we see a variety of only loosely connected cultures of conservation. Different countries are locked into different situations, each of which has quite specific implications for the shape and size of the national market for insulation. In conclusion, the same ever-so-simple technology, a technology that has broadly similar and undoubtedly significant environmental benefits, nonetheless faces an extraordinarily different future across Europe. Accepting this analysis, it is difficult, perhaps impossible, to place much faith in abstract models of technological diffusion or in estimates of future carbon dioxide emissions, which presume the steady adoption of proven energy-saving technologies.

Taking a positive view, the development of practical and realistic strategies for energy conservation depends upon an understanding of contemporary configurations of government, industry and consumer interest and of what these imply for the future. These relationships are not static and each new arrangement opens up different opportunities for the active promotion of insulation. For example, the growing international influence of insulation manufacturers creates new possibilities at the level of European regulation. More locally, the UK government's increasingly open style of consultation[21] has made it possible to debate and discuss the long-term future of building regulation (and thermal performance) with relevant manufacturers and trade associations. As these examples suggest, new institutional relationships have the potential to actively foster cultures of conservation and re-shape the contexts in which installers and manufacturers operate.

This chapter has examined the development and diffusion of a simple technical fix. In northern Europe better insulation represents a technically proven, economically beneficial means of saving energy. However, the error has been to see this fixing as an autonomous strategy with a life of its own, independent of any social or cultural setting. Whatever else, we have demonstrated the highly contextual nature of insulating practices. Although the material might be the same, the routes through which it has found its way into the walls and roofs of the European building stock are immensely varied. This observation has practical implications. Although governments, building researchers and environmental activists have the common aim of encouraging the use of this standardized technology, promotional strategies will only be effective to the extent that they take account of, and exploit, localized opportunities for change generated within different cultures of conservation.

6 Organizing design: housing and energy efficiency

The previous chapter suggested that conventional explanations fail to take account of the social complexity of technical change. The appropriation of energy-saving technologies is not inevitable. Nor are patterns of diffusion simply determined by need and demands relating to climatic considerations or political enthusiasms. Techno-economic models of choice and change have the further effect of highlighting, perhaps exaggerating, the role of individual decision-makers. As we have already seen (in Chapter 4), these ideas have their roots in a linear theory of innovation and an individualistic conceptualization of rational decision-making. Knowledge is the medium that spans these two tracks of thought. The underlying expectation is that greater knowledge of the potential for energy conservation will lead to energy-saving action. Hence, also, the expectation that better informed individuals will make better informed, more rational, and consequently more energy-efficient choices. Woven together, such ideas have provided the rationale for any number of promotional initiatives and demonstration programmes.

In this chapter, which is based on a study of UK housing, we re-examine the notion that the implementation of energy-related research depends upon the knowledge of key decision-makers. In addressing this issue, we also consider the relationship between knowledge and practice. Is it the case that better informed individuals do in fact act in more energy-conscious ways and, if not, what other accounts might we offer? One other account is, of course, that such individuals are prevented from acting rationally (at least with respect to energy efficiency) because of some non-technical barrier, such as the division of interests between landlord and tenant, the professional fee structure that rewards investment in expensive heating and cooling equipment, and so on. Chapter 7 explores the barrier concept in more depth, showing how energy-related practices in commercial office buildings relate to the balance of interests between landlord and tenant. But first we focus on theories of decision-making that guide the packaging and marketing of energy-related research.

Best practice and decision-making

There are, of course, many different types of science-based knowledge and not all are expected to feed, directly, into the worlds of practice. Some forms of technological development are adapted, managed and mediated by manufacturers. Other kinds of knowledge are highly esoteric. But in thinking about the broad base of the 'energy problem' much of the relevant technical knowledge is relatively straightforward. This is especially so with respect to 'simple' building types such as the house. Of course, houses come in many shapes and sizes, but, at least in technical terms, the energy-related problems and opportunities are generally limited. There is only so much that can be done to improve the energy standards of homes designed for the mass market. Renovation is another matter, but again the range of options and opportunities is often restricted. This is one reason why there have been so many 'best practice' case studies and so many demonstrations of simple ways to improve the efficiency of the housing stock.[22] Such initiatives view houses as interchangeable technical products, which are broadly comparable in terms of function, form, specification and design. Appropriate energy-related measures are similarly standardized and readily available. In short, the technology exists, the knowledge is there, it is easy to demonstrate and show, and if adopted by all those involved in producing and renovating homes, it would lead to a really significant drop in energy consumption. Cast in this way, the challenge is one of 'getting the message across' to relevant decision-makers.

Government-sponsored promotional programmes have become more sophisticated over the years and there is growing awareness of the need to differentiate between audiences and to provide tailored packages of knowledge for each. In the UK, for instance, the Energy Efficiency Office's Best Practice Programme was designed around a complex matrix in which types of decision-makers were ranged along one dimension and types of knowledge along the other. Specific promotional 'products' were then designed to suit the requirements of different cells in this elaborate marketing scheme. In some instances, the 'same' technical knowledge was promoted but with a different spin depending on the intended recipient. These strategies reinforce the view that there is a core of purely technical knowledge that must then be marketed, as sensitively as possible, to key decision-makers. The sensitivity has undoubtedly increased, yet the underlying model of decision-making has remained more or less the same.

In the case of housing, there is a limited repertoire of energy-related messages to be got across. The means of doing so varies from one context to another. For example, a number of UK initiatives have focused on volume housebuilders in the hope that other smaller companies will be inspired to follow the energy-related practices of market leaders. Likewise, the chief executives of Local Authorities have been the targets of rather diffuse 'awareness raising' campaigns in the hope that once enthused by the cause of energy efficiency, they will encourage their technical officers to seek further advice and guidance on this issue. Both examples tap into an implicitly hierarchical theory of knowledge and influence. Once the key people are persuaded, others will follow and, once enlisted, key decision-makers can, by definition, make key decisions that affect the practices of others.

This is only one way of thinking about organizational change and, as we argue below, it stops short of capturing the routine contexts and circumstances that frame, and in some cases determine, the range of options between which decision-makers decide. Knowledge is often but not always a pre-condition for energy-saving action, yet the structuring of choice and the capacity to act is ultimately what makes a difference to its translation into practice. The idea that there are key and less key decision-makers goes some way towards recognizing differential capacities for action (although the notions that effective power resides with chief executives or that volume housebuilders set trends that other completely different sorts of organizations follow are both contested). However, the problem with this approach is that the focus remains on decision-making rather than on the structuring of choice.

If we turn our attention to the structuring of choice, we have to revise the view that housing offers a relatively standardized array of technological opportunities for energy efficiency. In some abstract sense, this is, of course the case. But if we are to consider opportunities for energy efficiency in the real world, housing does not represent a uniform environment in which and to which a standardized suite of proven methods and measures might be applied. Re-cast as a composite socio-technical enterprise, housing is clearly not homogenous. Socially realistic and socially viable opportunities for even such simple measures as cavity wall insulation or double glazing are correspondingly diverse. And, as we show in this chapter, this has really very little to do with key decision-makers, the extent of their apathy or the depth of their ignorance.

The point becomes clear if we imagine an individual designer equipped with a stock of energy-related expertise, perhaps acquired through careful reading of case studies and demonstration projects. At one point in her career, this designer is employed by a private sector housebuilder. Her energy-related practices are those that mesh with this environment, never mind how much more she knows and how much more could (in some purely technical sense) be done. When she changes jobs and takes up a position in a Housing Association, a different fraction of her expertise comes into play. She has no more or less knowledge than before but her decisions differ dramatically.

To some extent this is again explicable in technical terms. For one thing, she is likely to confront different sorts of choices, working on new developments in the first job and maybe a range of renovation projects in her second role. As the most recent energy supplement to the English House Condition Survey (dating from 1986, published in 1991) notes, 'compared to only one in seven of owner occupiers, a third of both Local Authority and Housing Association tenants and over half of all private tenants had accommodation that was relatively inefficient' (Department of the Environment 1991: 5). We note, in passing, that the state of the existing stock is to some extent the consequence of past patterns of decision-making, and as the sweep of trends indicates, these 'choices' are shaped by collective social contexts and conditions, not by the personal preferences of those involved. To continue the story, the differential range of technical problems is only part of the reason why the application of knowledge differs from one organizational context to another.

Contexts of action

As the following sections show, there are strikingly different priorities at play. Interviews with managers, developers, designers, renewal and maintenance officers working in Housing Associations, Local Authorities and private sector housebuilding allowed us to investigate these in some detail. Even within one setting, for instance, that of a private sector housebuilder, people employed in the marketing department were unlikely to share the same aspirations as those in the technical or accounting divisions. The divide between housing sectors was still more dramatic. This is significant for the realistic, as opposed to the merely technical, potential for energy conservation is structured accordingly. There are positive opportunities in each case, but they lie in different places. In order to know where those places are, we need to understand how these sectors are organized and what this means for 'normal' standards of energy efficiency.

Shifting associations

Traditionally concerned with the needs of 'vulnerable' social groups, Housing Associations have experienced a series of quite traumatic changes since the 1974 housing act (Malpass and Means 1993). The Housing Association movement, which has grown from a patchwork of typically small, often co-operative organizations, has a reputation for its commitment to high quality, affordable housing.

The 1974 housing act, designed to stimulate housing provision, introduced a supply-side subsidy known as the Housing Association grant (HAG), the purpose of which was to encourage the development and improvement of controlled rent accommodation (National Federation of Housing Associations 1985). Housing Associations were able to use this subsidy to produce new homes with relatively generous standards of space and specification. The grant covered the majority of building costs and permitted developments that were, as a matter of pride, designed to stand apart from similar Local Authority schemes. The combination of a commitment to meeting social need and an amenable financial regime provided a positive context for energy-efficient innovation. Putting these elements together, the National Federation of Housing Associations' (NFHA) energy conservation working party issued a 'manifesto' for affordable warmth (1988). This document encouraged Housing Associations to raise energy standards in order to provide homes that tenants could afford to heat. The logic was one of social responsibility. On the same basis, the working party tried to persuade the Housing Corporation, which monitored proposed schemes and managed the Housing Association grant, to prescribe energy-related standards. This effort failed in part because the Housing Corporation's role was changing in line with a new approach to housing development.

The Conservative government's commitment to home-ownership through the 1980s had a dual impact on Housing Associations. They became the main focus of government housing policy while also being subject to rapid financial deregulation (Langstaff 1992). Loosened from the control of the Housing Corporation, the statutory body governing development activities, Housing Associations were obliged to experiment with commercial approaches to new construction. As part of this trend, the scale of HAG funding began to fall (Bromley 1993). Whereas Housing Associations had received almost 100 percent of the total costs of new development from government subsidy, this figure dropped to below 50 per cent. The shortfall had to be found from either internal or private financial sources.

The economic umbrella under which Housing Associations used to shelter has steadily folded. Associations that wanted to maintain a role as the developers of new housing have been rather suddenly pitched towards the private sector (Kennedy 1993). This has had profound organizational, technical and cultural consequences, often leading to the wholesale remodelling of priorities and practices (Kelly 1994). The charitable framework of provision has, for example, given way to a new regime in which associations grapple with the complexities of the commercial world. In this context, energy-related decision-making takes its place alongside, and often below, a whole host of other priorities.

The Housing Association movement has always had a split organizational culture. Large Housing Associations have co-existed alongside tiny, specialized collectives, both satisfying a great variety of housing need. This is a complex world and it is something of an oversimplification to distinguish between large and small associations. Nonetheless, the distinction is useful, especially because the divide has widened with larger associations actively pursuing development projects on a nation-wide scale. When they do get involved in new building, smaller more 'traditional' associations do so on a very different financial and commercial basis. Although following different paths, all Housing Associations have had to invent new ways of funding their activities. Examples include forms of consortium development where several Housing Associations work together or co-operate with private developers to allow construction on a grander scale from that they might have previously imagined possible (Graham 1992). The result has been the emergence of a development super-league in which the bigger developing organizations compete with one another for the reduced housing subsidies still available. The changes outlined above have also served to shift the focus from inner-city regeneration to the more predictable and profitable green-field sites favoured by volume housebuilders. 'Capacity-expansion' is the aim for the larger associations. Overtaken by this ambition, historical commitments to a local area or population are often rendered meaningless. Summing up the situation, a Local Authority housing manager expressed the view that:

Their [the Housing Associations] principal interest at this stage is ensuring that they [new homes] are soundly built with no maintenance costs. They are trying to play as safe as possible by building as effectively as possible and soundly as possible. They are saving money by reducing space and ignoring the extra running costs of the building.

Meanwhile, smaller Housing Associations fear for their continued existence. Often anxious about the predatory tendencies of larger rivals, they have been tempted to protect themselves by preserving financial reserves. One typical strategy is to stockpile financial resources while striving to keep rents low to ensure the loyalty and social commitment of the tenants. Such Housing Associations are content to provide a limited, but effective, service to specific local communities, but are reluctant to over-stretch themselves, even in the cause of energy efficiency. As a housing officer from one of the smaller associations explained:

Energy conservation is one of those things that you do the best that you can when you can. I'd like to think that we can do more than that to be honest but it is not of prime importance provided we meet the criteria and the building regulations.

This split culture is implicitly sanctioned by the Housing Corporation. Committed to maximizing development activity, the Housing Corporation subsidizes Housing Associations able to provide the lowest cost per unit bids. With the Housing Corporation emphasizing quantity rather than quality, larger associations have sought to maximize their potential for growth by adopting the methods and techniques of speculative builders (Stewart 1993). Collaboration with private volume house builders has become popular. This, together with the need to pay close attention to costs, often through the development of standard specifications, has led to the successive edging out of 'superfluous' design details (Nelson 1993). One result is a new uniformity of style and standard, and with it an erosion of the distinction between private and public development. This creates a very different context for energy innovation. The move towards standardization gives individual designers little scope to influence energy measures and militates against opportunistic improvements and, in particular, developments. Although the costs of including higher levels of insulation or more efficient heating systems might be squeezed within individual development budgets, their inclusion in a standard design specification would mean greater financial commitment all round. As Housing Associations move into an ever more commercial environment, 'their' technical standards are increasingly set by the private construction companies with whom they work. According to one development manager, minimum levels of energy efficiency are becoming the norm.

> As you have probably found elsewhere, the issue of energy efficiency doesn't rate highly. There are so many constraints on getting things built that by the time you have sorted out funding and planning and you've found the piece of land and you've negotiated and you have gone through all the hoops and jumped all the hurdles, just to get anything on site is a major achievement.

Housing Associations' fears and desires clearly influence their technical choices. Nonetheless, practitioners working within both large and small associations were united in the view that commercial pressures have led to reductions in space standards and levels of energy efficiency. There is some evidence to suggest that space standards are 'slipping dramatically' (Stearn 1994).[23] Rather than setting and maintaining standards, the government is urging the Housing Association movement to compete by minimizing costs per unit. In this context, the compromise of previously preferred design principles seems inevitable.

This commercial logic is accelerated by the Housing Corporation, which has reduced budgets for individual developments and moved away from a prescriptive role in quality control (Karn and Sheridan 1994). This is reflected in the production of advisory 'scheme development standards' (Clowes 1993). These 'advisory standards' are a world apart from the previous regime of rigorous inspection and mandatory requirements. Additional energy-efficiency measures are merely recommended and there is no obvious motivation to enhance energy performance beyond the minimum laid down by the building regulations.

This change of emphasis is not quite overwhelming. Examples of high-quality design and increasing energy standards do exist, often within niche markets. Some associations are still wedded to energy efficient design as part of a broader commitment to the provision of affordable housing. There are other signs of a spirit of resistance to the downward spiralling of standards. Certain associations are, for example, funding improvements in energy efficiency by raising rents, the burden of which is carried by the government through housing benefits (Chaudhary 1993). The result is a more efficient housing stock for the association and smaller fuel bills for low-income inhabitants, both achieved by a somewhat circuitous route.

The more general picture is, however, one in which the vision of 'warm homes for all' has faded in the drive for growth, and as Housing Associations' roles and ambitions have evolved in response to the changing political and financial climate. One obvious consequence is that these shifting associations of energy and housing have generally made it more difficult for knowledgeable and capable designers to push through energy-related innovations. No amount of personal commitment or enthusiasm can make up for the changing context of decision-making or bring back the days of 100 per cent HAG funding. More positively, the changes we have outlined here have different implications for larger and smaller associations. As a result, the logic and rationale for energy conservation and the means by which it might be achieved have become more diverse. Housing Associations involved in major development programmes might, for instance, adopt good energy-saving practice not for its own sake but because of a long term interest in the cost of maintenance or as a side-effect of standardizing some un-related aspect of construction. Smaller associations might adopt similar measures but for very different reasons.

Local pressures

The move away from government investment in the public sector has had a similarly powerful, but very different, effect on Local Authorities' efforts to raise the energy standards of their housing stock. Again a little history is helpful. The desperate need for housing following World War II was partly satisfied by Local Authority provision. A large stock of new housing was quickly created in areas of social and physical devastation (Colquhoun and Fauset 1991). Although standards of space and heating were not as high as in private developments, they were often an immense improvement on the slums they replaced.[24] Some Local Authorities made great progress in developing a condensation-free, thermally efficient housing stock (Goodchild and Furbey 1986).

As with Housing Associations, political shifts over the last decade or two have fundamentally altered patterns of Local Authority housing provision. Critically, the political desire to create a nation of homeowners led to a dramatic decline in investment in new Local Authority housing. Under the 'right to buy' scheme, the 'best' and in the context of our discussion, often the most thermally efficient, stock soon disappeared. As one civil servant commented:

> I can tell you I have got a new reason that I was given by a Local Authority last week, they will not spend on energy efficiency because their dwellings will be subject to the right to buy, the better they make them the more easily they sell. This has proved to be the case in another authority whose energy rating on average is lower this year than last because they have sold some of the dwellings that were improved.

This has left Local Authorities with a confused role and a particular set of problems. Rather than shaping local housing strategies around the goal of increased public provision, housing departments instead contend with the management and improvement of a decaying building stock. According to the English House Condition Survey, Local Authority housing has particularly acute energy problems; 81 per cent of all cavity walls are unfilled, 57 per cent of lofts have less than 100 mm of insulation, 79 per cent of dwellings have only single glazing and 22 per cent have no central heating (Department of the Environment 1991).

The political and economic upheavals that have shaken the Housing Association movement have similarly reduced and changed the form of state aid for council housing. Local Authorities seek funding for development and rehabilitation programmes through the housing investment programme (HIP). As with HAG (see above), HIP used to provide for almost 100 per cent of costs, allowing for cyclical improvement schemes within which thermal standards could be improved through packages of retrofit insulation (including cavity wall insulation) and more efficient heating systems. This has all changed. HIP contributions declined from 90 per cent to below 60 per cent of improvement costs (Bromley 1993). Although the government claims that there is more money available for renovation grants, Local Authorities are expected to find the additional (40 per cent) contribution from local sources.

With an overall decrease in central government provision, subsidies are frequently targeted on housing in the most advanced stages of dilapidation. Only the worst cases are receiving attention. This soaks up funds more rapidly and often to less effect, at least from the perspective of energy conservation. Apparently benevolent compensation rules designed to benefit occupiers displaced because of demolition are also potentially counter-productive. Prior to the 1990 Housing Act, owners only received compensation for the value of the site. Now full relocation costs must be paid. This often makes demolition uneconomical, leaving even more dilapidated housing to be patched up, or sold on to private contractors.

The style of funding is also relevant. Government conservation initiatives are increasingly competitive, requiring Local Authorities to submit bids that are evaluated according to varying and often changing criteria. Significant sums of money are filtered through these schemes, but it requires concerted effort to learn the rules of the game and establish the necessary contacts and partnerships. The Energy Savings Trust, set up by the government and major energy companies in response to the 1992 Earth Summit, provides advice, guidance and a range of services in support of the 1995 Homes Energy Conservation Act and the Home Energy Efficiency Scheme. However, the Trust has a tendency to concentrate on highly visible schemes, promoting initiatives such as the HECAction competition. These are very significant for those who win but make little impression on the scale of problem which Local Authorities face.

Our respondents also argued that strategic maintenance plans were constantly thwarted by unexpected policy shifts. Although environmental concerns provide an increasingly important rationale for government involvement, decreasing funds for capital programmes mean that real ingenuity is required to secure steady streams of funding for long-term programmes of energy-related improvement. As a result, planned programmes are giving way to targeted bursts of 'emergency' maintenance initiatives, or sudden flurries of action following a windfall of competitive success. As a housing renewal officer told us, under these economic conditions sustained consideration of energy conservation is difficult.

> Energy conservation ... in all probability is going to get sidelined and we are going to have to say well alright we'll stick with the old single glazed window, they've got another 30 years life in them.

In this context, the appearance of energy-conservation criteria in HIP applications is causing some consternation (DoE 1994b). The Home Energy Conservation Act also requires every Local Authority to prepare detailed plans as to how it would cut energy consumption in its homes by 30 per cent. The problem here is not only one of conflicting financial priorities but also of changing culture. Local Authorities are now being encouraged to develop computerized stock profiles, utilizing energy-rating systems such as the standard assessment procedure 'SAP' to survey and map the energy characteristics of their properties (DoE

1994b). This provides a base of data against which to compare and assess refurbishment proposals. It also means that measures and pro-grammes can be evaluated in terms of costs, benefits, and contributions to carbon dioxide abatement targets. This data gathering is all well and good, but in many authorities, housing management services have depended on the long-term familiarity and wisdom of experienced per-sonnel. Re-orienting management practices takes time and at least some of those we interviewed were concerned that data gathering soaked up resources better spent on energy conservation, and merely revealed the enormity of the problem. Meanwhile, a select few authori-ties are leading the way. Familiar with the language of energy-rating, they have swept the board in terms of attracting additional government subsidies and have gained an edge in HIP bids. This is again generat-ing a split culture of conservation in which 'innovative' Local Authori-ties are rewarded for significant achievements whereas the difficulties others have are overlooked. As a maintenance officer explained:

> What we had to do was present them with a programme of what we are going to do and we also have to give them statistics on how we have performed in the past, what we have done in new build – which is nothing – or what we've done in repairs and what have you. They ask for a lot of statistics and now for the first time they are saying can you tell us as a statistic what the energy rating of your property is? We can't do that. Now this year it doesn't matter, but we think in future years governments are going to say, 'well, if you don't know what the energy ratings of your properties is you are not a very good authority so we are not going to give you any money.' We will be penalized.

There are other inequalities associated with the way in which housing improvement is planned and managed. Although the Department of Environment Transport and the Regions encourages long-term plan-ning, HIP awards are made annually. This means that Local Authori-ties must prepare future initiatives without knowing what resources they can expect more than a year in advance. These uncertainties are compounded by the necessity of dealing with widespread social pro-blems. In terms of housing management decisions to improve particu-lar neighbourhoods, these are often influenced by political expediency as well as by technical need. Organizational transformations introduce yet more layers of uncertainty and potential confusion. For example, the division of departments into competing directorates can alter traditions of inter-departmental communication with the result that council-wide energy practices may not be linked to housing improvement strategies.

In sum, many Local Authority housing practitioners acknowledge that energy-efficiency initiatives can play a key role in revitalizing dilapidated housing, but point to the huge tensions that exist given the scale of the task, budgetary restraint and organizational uncertainty. Although government has recognized the contribution that the thermal improvement of the Local Authority stock could make to the reduction of carbon dioxide emissions, housing practitioners seem to face an unending stream of problems (Friends of the Earth 1994). As one told us:

> You've got to meet so many other criteria, the planning, access all that sort of crap plus all the local government and government relationships and you overcome all the constraints government puts on you. So for a Local Authority to do anything is very, very hard and the culture that Local Authorities are now working in is one of resignation, one of insecurity, one of low morale. I can imagine a considerable reluctance to stick your head over the parapet. At the same time there is a feeling that the government isn't committed to energy conservation.

Bridging the 'efficiency gap' is evidently more than a matter of improving the flow of knowledge and information. Promoting a government green charter will not in itself change energy-related practices within Local Authorities for these are already strongly determined by patterns and styles of government funding and by the sheer scale of the challenge involved in managing a physically deteriorating stock. In the absence of new build, design and development strategies are geared to worst-case renovation in which energy efficiency takes a backseat to more pressing social problems. Individual entrepreneurs and enthusiasts can still make a difference. In fact, growing reliance upon competitive modes of funding enhances their chances of doing so. However, competitions have relatively few winners and a large number of losers. The bigger picture therefore remains one in which opportunities for energy efficiency relate more to the future and funding of local government than to the knowledge and motivation of key decision-makers.

Private dilemmas

We now turn to the private sector. Commercial housebuilders are not subject to the same sea changes or political pressures as Local

Authorities or Housing Associations, but nor do they inhabit a stable world. Instead, private housebuilders negotiate a fluctuating market in which the value of land plays an important part. The economics of speculative, volume construction favours standardization, and the development of pattern-book house types that can be superficially modified to suit the tastes of individual buyers. Housebuilders believe that it is important to spend money on bathrooms, kitchens and possibly double-glazing in order to attract customers (Goodchild and Furbey 1986). Extra insulation, like other elements of the structure, invisible to the end consumer, is only rarely considered to be a marketing feature. This housebuilder's experience was typical.

> they know that double glazing is something to do with energy saving and they can actually see it, but if you try to explain to them that within their wall they've got 100 mm insulation, [or] no insulation but it's got a highly insulated block or something like that, they are not interested because they can't see it.

For this reason, amongst others, the energy standards of private housing rarely exceed the thermal requirements of the building regulations. Design specifications are instead driven by the building industry's sense of market demand. How these senses are acquired remains something of a mystery but 'gut feeling' is an important consideration. Attempts to exploit 'green concerns' by developing low-energy houses that cost a little more than a 'standard' alternative have not been judged a success. For commercial builders, including the one quoted below, the acid test is whether, and at what rate, the houses sell (Ball 1983).

> people do not ask if their house is highly insulated or very rarely will they ask whether it's double glazed and they will ask whether it's got this and that and does it include fitted carpets and does it include cookers etc. Etc., 'ooh that's got a hob, is it a fan hob?' 'Yes it is.' And they will accept those sorts of things but very rarely will they actually come to you and say to you what is the insulating value of the walls.

The fact that a group of energy-efficient homes failed to sell faster than similar 'conventional' properties confirms the view that energy-related features are generally ignored. With these gut feelings in place, builders have little incentive to deliberately promote energy-efficient homes. To give one specific example, the idea of improving levels of insulation in order to introduce smaller heating systems makes perfect technical and sometimes economic sense (if the cost savings on heating are invested in insulation). However, housebuilders respect and reinforce a psychology of heating that equates radiator size to comfort levels.[25] Consumers are thought to want systems that can respond quickly and still have capacity in reserve. In order to protect themselves from blame, heating sub-contractors will also 'over specify', resulting in the selection of boilers that consistently operate below optimum efficiency. Often looking for cost savings through bulk buying, builders are tempted to fit the same heating system across the full range of house designs. Such inefficient decisions make sense given the level of commercial risk attached to the first whiff of consumer dissatisfaction and given the 'real' economics of the market for boilers.

Housing Associations and Local Authorities have a long-term interest in maintaining efficient properties because money spent on unnecessarily high heating bills means that less is available for the rent. Private builders do not share these concerns. While commercial developers have little interest in the economic performance of the systems they install, they may be held liable for their technical failure. Private builders are therefore reluctant to take what they think of as 'risks', even if these promise significant benefits in the form of lower running costs and greater energy efficiency.

Housebuilders' resistance to fully filling cavity walls illustrates this point. Despite government-funded research into the reliability of cavity wall insulation the National House Building Council (NHBC) has been suspicious. These anxieties relate to the risk of rain penetration and subsequent dampness for which builders might be liable.[26] Still quoting insurance claims resulting from the storms of 1989, housebuilders and their representatives lobbied hard to ensure that the 1995 revisions to Part L (the thermal part) of the building regulations did not put them under further pressure to fully fill cavity walls (Chevin 1991). What was ostensibly a debate about technical risk was in fact coloured by concerns about risk of another kind, that is commercial risk and risk to the reputation of individual builders and the image of new houses in general. One interviewee explains the difference between his own position and that of a Housing Association.

> They [Housing Associations] are not selling them [that is houses]. If they get moisture coming through all they do is send their maintenance people along and they repair it. If we get moisture coming through and it gets on one of our sites it can make the

difference between selling houses or not. Once you get a bad name it's commercial suicide. Barratts on their timber frames just because World in Action went along and looked at that site and saw polythene sheets being split they lost a fortune, the share price went down and people stopped buying them.

In this context, the building regulations are all important. Builders generally argue that the only way to persuade them to put science into practice and adopt higher standards of energy efficiency is to force them to do so. Such forcing is easier said than done. Regulating is a complicated process in its own right and one that is influenced by a whole variety of interests and priorities. Housebuilders have typically opposed any increase in mandatory standards, usually on the grounds that the cost of new homes will increase and jobs will be lost (Goodchild and Furbey 1986; Rydin 1986). The current move towards performance-based rather than prescriptive standards, the adoption of a longer term view of the consultation process, and the inclusion of energy rating within the regulatory framework serve to confuse what seemed to be a simple opposition between government and industry (Shove and Raman 1996). Yet it is clear that regulation cannot take place far in advance of normal industry practice and that there is no intention, at least in the UK, of using the building regulations as a means of forcing energy-related innovation or of imposing the hard won lessons of building science.

According to housebuilders' own accounts, innovation takes place within, and because of, highly stringent commercial demands. Developing this theme, they argue that there is no reason why they should build 'green' houses if consumers do not want to buy them. Equally, there is nothing to prevent them adopting higher standards if that makes sense from a commercial perspective. Environmental innovation is not entirely out of the question. There are, for example, signs that the idea of 'homes for investing' is perhaps selectively giving way to an alternative view of 'homes for nesting'. The pattern in which households move on every five years or so in order to capitalize on their initial investment and accommodate growing families appears to be changing. With property proving to be a less predictable investment, and with a fast-growing number of smaller households, people are staying put for longer. This has implications for how the home is viewed. Longer-term perspectives appear to generate a new interest in the quality of construction. In addition, measures that have uneconomic pay-back periods for those who expect to move on may represent sensible investment strategies for committed 'nesters'. Of course these trends still have to be translated into 'gut feelings' if they are to make a difference to new house building practice. The more general point is that the energy standards adopted by private sector housebuilders are malleable, but are rather more sensitive to beliefs about market trends than to the results of building science.

Context, choice and change

In this final section we return to the question of when and why individual decision-makers put proven technologies into practice. Where the measures are simple and cost effective, and where the technologies are familiar – but still not adopted or used – it is easy to point the finger at the individuals involved. Hence the view that the challenge of energy efficiency is as much one of overcoming apathy as of providing more and better technical understanding. In taking this line, commentators draw upon a widespread image of the construction industry as one dominated by people who suffer from 'an excessive lack of moral scruples or too much elitist insensitivity' (Ball 1988: 32). As Michael Ball observes, key players have been on the receiving end of blame and condemnation for different reasons at different times, witness the 'get-rich-quick' speculator blamed for poor quality construction (Colclough 1965) or the planners held responsible for the council housing of the 1960s and 1970s (Ravetz 1980). Likewise, a number of architects have fallen prey to corruption (Gillard and Tomkinson 1980) and, most common of all, contractors have been accused of cowboy practices (Ball 1988). But as Ball points out:

> while cases of professional ignorance and insensitivity, and individual malpractice can always be found, to explain structural problems by 'individual failings' is both poor theory and false moralizing. What conditions enable such practices to exist?
>
> (Ball 1988: 33).

For Ball, explaining 'failure' in terms of individual action confuses the symptoms with causes. He suggests that it is more appropriate to view individuals in their 'social context rather than isolate them and exaggerate their individual characteristics, as the social-failings approach tends to do' (Ball 1988: 33). Making a similar point about domestic consumers, Hinchcliffe argues that 'energy studies need to stop assuming the characteristics of the individual consumer, and blaming people for not acting in the way set out in economic brochures' (Hinchcliffe 1995: 94).

What do these points imply for individualistic theories of decision-making such as those that have influenced government efforts to promote the efficient use of energy? At first sight, they appear to point the 'blame' elsewhere. Those initially identified as the key decision-makers turn out to be the wrong people (Guy 1999). Instead of looking at the building industry, or at Local Authorities or Housing Associations we should, perhaps, search out the regulators, or those responsible for re-formulating local government funding, or for failing to exert sufficient influence over the actions and practices of designers and others who make decisions on the ground.

Another interpretation, and one that is more convincing in the light of the material presented here, is that the search for key decision-makers is doomed to failure. Such a mission implies that it is possible to define and attribute agency and responsibility in some relatively simple fashion. The cases we have described tell another story. Yes, decisions are made and yes, they are made by real individuals. But that should not lead us into the trap of thinking that the course of energy efficiency is straightforwardly determined by individual decision-makers. Although practitioners do make strategic energy-related choices, their capacities for action are socially structured. As we have seen, what happens in practice is the momentary outcome of a temporally and situationally specific combination of conditions, circumstances and priorities. This argues for a different way of seeing energy-saving action. What is required is a means of understanding how such conditions and circumstances arise and how they frame interpretations of sensible and worthwhile courses of action. In theoretical terms it makes no sense to abstract actors from the contexts in which they are situated or to talk of decision-making without also asking about the framing of the decisions themselves.

But this is not just a theoretical point. It is also one that has far reaching practical implications. The National Audit Office's claim that there is a 'range of practical and cost-effective measures which have the potential to save up to 20 per cent of the energy consumption of buildings' (1994) echoes hollowly when extracted from the organizational realities of the different housing worlds we have described. In what sense does this potential really exist? As we have seen, what counts as 'practical' and 'cost effective' depends upon the time and place of housing development. If the first 'mistake' is to view housing professionals as isolated decision-makers, the second is to assume that because they deal with a similar product they inhabit similar worlds of practicality and cost.

This brings us to a difficult point. If the application of energy-related expertise is so contingent and so subject to change, is there anything that can be done to actively promote technology transfer? If we argue that model of individual decision-making is deeply flawed, is there anything to offer in its place?

Some of the ideas referred to in Chapter 4 are useful here, but again in theoretical rather than practical terms. For instance, concepts of path dependency, and of socio-technical regimes help make sense of the trajectories of change described above. However, these ideas are of little use to those wanting to engineer new relationships on the ground or to actively promote the cause of energy efficiency. Although we do not aim to offer definitive prescriptions for policy or practice, we can draw some conclusions. First, this chapter has outlined the very different situations in which Housing Associations, Local Authorities and private sector housebuilders operate. It has also hinted at the existence of pockets, niches and differences within these sectors. For example, some Housing Associations viewed the provision of energy-efficient homes as an essential, not an optional, role. Some Local Authorities made and were able to implement strategic decisions to play the energy game and compete for funding in that area. Finally, some private sector builders have spotted, and gone out of their way to exploit niche markets, for example, providing more energy-efficient homes for the elderly. Others have inadvertently adopted energy-saving materials and processes for reasons of economic efficiency (Egan 1998). Each of these micro-situations lends itself to the practical application of one or another form of energy-related expertise and therein lies the lesson. Taking a more modest approach to the promotion of energy efficiency, it is important to acknowledge that measures and technologies are not directly transferred, but are picked up and utilized as and when they make sense in particular circumstances. Recognizing this, policy makers and technical researchers should be able to identify and help exploit organizationally as well as technically viable opportunities for change. In this way, they can work with the grain of current practice, not against it. The only trouble is that current practice does not stay still. As a result, and as we show in the next chapter, the potential for energy conservation is a highly dynamic concept.

7 Developing interests: office buildings and barriers

In the previous chapter we examined the idea that energy-related choices reflected the technical knowledge and environmental commitment of key decision-makers. Comparing the contexts in which housing practitioners operate, we showed how standards of energy efficiency were framed by different configurations of purpose, possibility and social potential. This led us to develop an alternative model of choice in which the 'decision-maker' is treated as 'a person embedded in a network of social relations that limits and controls the technological choices that she or he is capable of making' (Cowan 1987: 262). In this chapter we focus on the issue of change. In the process we re-examine the idea that social barriers impede technical progress by showing how the social contexts of choice evolve. Focusing on the commercial office sector, this chapter explores the ways in which the dynamics of the market, investment conventions, varying lease arrangements, new approaches to building management, and the changing status of users underpin energy-related developments.

Examining the history of the commercial property development in the UK, we argue that design and development is a relational activity involving multiple actors. Associated patterns of decision-making may be collaborative or highly contested, but they always takes place under conditions in which some actors have more power than others. We show how the social organization of property varies geographically and culturally by examining other European, and in particular, French contexts of development. The purpose of this comparative review is to show how such relationships, including that between landlord and tenant, make a difference to energy-saving action. Examining differences between the UK and France, and following the changing configuration of British property interests over two decades, we illustrate the interpretive flexibility of energy efficiency.

By following office development in Britain we see how interest in energy efficiency ebbs and flows with the market. The case of France, a development landscape defined by a different patchwork of legislation, practice, codes and conventions, illustrates an alternative pathway of development. Semi-structured interviews with developers, investors, property agents and occupiers on both sides of the Channel provide the raw material for the following narratives of development practice.

These rather detailed stories have important implications for the analysis of non-technical barriers to energy efficiency. In this context, the classic 'barrier' relates to the different interests of landlords and tenants. Government-funded studies of the potential for technical innovation in the office sector frequently highlight this problem. The approach taken by the authors of a 1988 market research study for the UK's Building Research Establishment Energy Conservation Support Unit[27] is typical. Their report provides a statistical portrait of the commercial building stock, also forecasting expected trends in development and future construction. Descriptions of the principal actors, their typical role in the development process and their attitudes to energy performance add flesh and colour to the numerical data. But at this point the social, organizational and economic complexities of the property world also become apparent. No single model of the development process conveniently presents itself – custom-build for owner–occupation, pre-let building for sale or retention to a known occupier, or speculative development for an uncertain market – all involve different relationships between owners, occupiers, agents, developers and investors. This bewildering complexity has to be simplified if key decision-makers are to be identified, informed, and persuaded to adopt more energy-efficient practices. The strategy, in this case as in others, is to develop a two-part classification that simply distinguishes between speculative development and that which is destined for owner occupation. Although most UK office building is speculative,[28] the split between the developers' interests and those of unknown future occupiers constitutes what is generally taken to be a 'barrier' to energy efficiency. Developers have little stake in reducing energy costs as they have no long-term interest in the buildings they produce. Once the property is let, they have no responsibility for the energy bills. Moreover, if developers are to be able to sell their properties on to risk-averse investors they must take care to avoid environmental (or any other kind of) innovation, which puts their investment at risk without also maxi-mizing long-term rental return. By contrast, the owner occupied sector is expected to be a more receptive target. Where the interests of developer and owner are closely linked it makes sense to consider operating costs over a longer time scale.

On the basis of analyses such as these, the advocates of energy effi-ciency have drawn up typologies of key decision-makers. The trouble is that these typologies represent static snapshots that fail to capture the complexity of the property business. The production of specula-tive and owner-occupied developments follows a similar sequence in which evaluation precedes site acquisition, design, construction and on-going facilities management. Advocates of energy efficiency again simplify this complexity, focusing on the design stage to the exclusion of all else. In practice, assessments of energy-related expenditure always take account of capital and recurrent costs, technical feasibility and the impact of such investment on rental income and letting performance. As such, decisions are rarely the province of one occupational group alone. Owners, developers, engineers, agents and occupiers all interpret and value energy efficiency in different ways and all seek to steer the design process accordingly (Guy 2000). As we show below, attempts to locate and influence key decision-makers are constantly foiled by the shifting configurations of power that shape the course of commercial property development.

The pressures are certainly dynamic. As the national economy speeds up and globalizes, commercial organizations find it harder to predict their need for office space (Lizieri 1991). In this context, it is important to develop commercially viable buildings that can be sold on or leased for a competitive rate in the event that they are no longer needed. The pre-requisite for generating a commercially accep-table rent (necessary to attract an investor) is a commercially accep-table specification, high enough to meet the standards of the most demanding occupier (Goobey 1992). Agents, who advise users and investors on the relationship between specification and rental levels, have a critical role in managing this supply driven approach to prop-erty valuation. Drawing on knowledge of recent property transactions, their advice to match or better the specification of competing develop-ments tends to encourage the homogenization of office design. This is interesting in its own right, but for our purposes, the critical point is that this focus on commercially 'acceptable' specifications applies as much to custom-built property as it does to speculative developments. Distinctions between the interests of owner-occupiers and speculative developers make little sense when we take account of the longer term and realize that owner-occupiers and speculative developers inhabit exactly the same commercial environment.

Rather than searching for key decision-makers, we try to show how the changing social organization of the property business frames the relative power of the different actors involved, and how this opens up and closes down the potential for energy efficiency. The following sections consider the recent history of property development in these terms.

Property relationships

The roots of contemporary property development in the UK lie in conservative post-war legislation designed to encourage real estate activity. The reduction of planning restrictions, the removal of taxation on development profit and the elimination of betterment levies transformed the commercial potential of property speculation (Marriot 1967). Construction came to be seen as a convenient means of rapidly increasing the value of land. In this way buildings represented little more than an economic symbol: a source of commercial value that barely related to the form, specification or eventual use of individual offices. Valuing the built environment became a simple function of the expected income (rent flow), minus the development costs (professional fees, land and development costs, etc.) multiplied by the expected 'yield' (assessment of risk) of the developer/investor (Goobey 1992). Many academic commentators have since typified the development process as a struggle by property developers and investors to extract as much 'exchange value' as possible from building construction with little regard to the eventual 'use value' of the building they provide (Luithlen 1994). In this way the production of the built environment is seen to be driven by profit-seeking property actors, operating in a cyclical economic context, driving flows of capital through the development, demolition, and re-development of rentable space (Harvey 1989). If we are to understand this process, we have to take note of events at the 'intersection between agencies, structures and contexts' where buildings are produced (Madanipour 1996: 137).

From this standpoint we suggest that commercial development in the early 1950s was shaped by a pressing need to accommodate an expanding white-collar sector, by legislative support and by the evident commercial potential of construction. This stimulated a new entrepreneurial spirit and an intensification of development activity. Spectacular packages of planning consent and finance led to the multiplication of urban office space. By the 1960s, supply began to catch up with demand. A surplus of space threatened rental levels and undermined the key criteria upon which profitable development equations depended. In response, the newly elected Labour government banned office development in London and re-introduced a betterment levy on development profit (Marriot 1967: 11). The resulting drop in activity solved the developers' dilemma and rental levels continued to rise. By 1970 a similar reversal in the political climate again coincided with a shift in the balance of supply and demand. The Conservative government repealed legislative restrictions, reduced interest rates and increased the money supply. As the economy grew and the service sector expanded further, development activity boomed once more (Cadman and Catalano 1983). The full story is recounted elsewhere (Marriott 1967) but even a brief review of post-war development highlights the importance of legislative change, expanding service sector needs and entrepreneurial acumen, all contributing to the creation of a commodified built environment largely dedicated to the production of exchange value.

Investing and exchanging

In the mind of the financial community, property became firmly associated with capital and income growth. Large investors, pension-funds and insurance companies were attracted by the return, and low level of long-term risk of property investment, by the need to balance their portfolio and by the 'psychic' value of a tangible, visible asset (Baum 1991). Investment in property was made particularly compelling by a legal framework that gave almost total security to the landlord. Before the war, leases often stretched for ninety-nine years with little or no rental increment. The onset of inflation following the war re-fashioned lease terms. Leases were progressively limited to twenty-five years, and five-yearly upward-only rent reviews were introduced. These binding leases placed liability for repair and maintenance with the lessee and were further subject to 'privaty', the acceptance of full liability should any later assignee default on rental payments or maintenance responsibilities (McIntosh and Sykes 1985). As the post-war boom was sustained, the presence of increasingly wealthy investors was more strongly felt. Unable to spend their substantial funds abroad because of exchange controls (until 1979) and with British equities performing comparatively poorly because of the weakness of the manufacturing sector, property became an irresistible long-term investment. Dissatisfied with merely providing fixed-interest loans, investors sought closer financial partnership with developers in order to take a greater slice of the profits (Marriot 1967). Taxation changes in the mid-1960s targeted property companies, but ignored rental incomes from buildings held as investments. This encouraged institutions to buy buildings outright (Cadman and Catalano 1983) and the balance of investment portfolios began to shift sharply in favour of property. With the property crash of 1974–6 following the earlier economic recession, development activity halted. Rents had fallen and with construction costs rising in line with inflation, property prices plummeted. This allowed pension and other investment funds to consolidate their position as 'investment barons' by picking up office space, still constituting a 'prime' investment, at bargain prices. By the early 1980s, around 83 per cent of commercial and industrial property investment was accounted for by insurance companies and pension funds (Cadman 1984). The result was the establishment of a remarkably uniform structure of commercial and institutional priorities that have continued to dominate and determine the course of subsequent developments in the commercial property sector.

The implications for energy efficiency have been significant. In order to safeguard investment, buildings have become increasingly standardized (Madanipour 1996). In the office sector, investment potential is not exclusively determined by location, location, and location, as is often claimed. Details of design and levels of specification are also important. Driven by the need to ensure the safety and return of their investment, which as we have shown above is based on the reliability and level of the rental income stream, investors preferred to maximize the long-term attractiveness of their buildings by demanding high levels of technical specification. Internal environmental control came to be seen as essential, and lighting levels and power and floor loadings were also designed to surpass the requirements of the most stringent potential occupier. These are the roots of what has become known as the 'institutional office specification'.

The environmental ramifications of this institutional grip on office development are two-fold. On the one hand the institutions' desire to create a reliable, long-term investment meant that, in principle, they would only fund or buy 'the best'. This has encouraged developers to emphasize the quality of construction, and in some sectors of the market, there has been no temptation to cut corners on construction costs. As a result, occupiers have been provided with a 'better' and often a more flexible building than they 'really' wanted. However, it is equally clear that the pursuit of the 'prime' was a consequence of developers' exclusive emphasis on the 'exchange value' of office space and that it bore little or no relation to 'use value', to the needs of potential occupiers or to environmental concerns.

The following discussion of air-conditioning highlights the implications of the 'institutional specification' for energy efficiency. There are a whole set of beliefs surrounding the need for air-conditioning in the development process. Greater flexibility of potential use, a higher standard of internal environment and a prestigious, more marketable product are just some of the institutional benefits. The views of one commercial development manager make this point explicit.

> I was asked by our group manager the other day why are we looking at providing air conditioning in this City building and my response was because the market demands it and we as speculative developers and as a commercial organization can't afford to buck the market, we can't tell it what it wants. If we were the sole developer in the UK and provided all products then we could say this is all you are getting and the tenant would have to accept it because he would have no other choice. We are very small fish in a big pond.

The economic and environmental consequences of this process are clear. Like for like, energy costs and carbon dioxide emissions tend to be at least 50 per cent higher in air-conditioned offices (Harris 1993a).[29] Employees, worried about their health, and employers, concerned about sick building syndrome, are often suspicious of air conditioning systems (Tyler 1991). However, the customary equation of 'prime' specifications with secure, peak rental income has made it very difficult for designers to opt for natural ventilation.

In defining the form and specification of an ideal type commercial investment, the institutions have had a controlling influence on the process of development (Goobey 1992). Agents continue to play a pivotal role here. In the unique position of advising all other actors on what the market offers and demands, they can consistently insist on prime specifications as a benchmark for yields and rental performance. As many funds lacked experience in property management they looked to established agents for advice (Goobey 1992). The message was simple. New developments had to maintain a parity of specification if comparable rental levels were to be generated (Plender 1982). As an architect commented:

> Our own experience in UK work is that many estate agents, developers and (behind them) institutional pension-fund managers

consistently act irresponsibly in denying experimentation and demanding inefficient building solutions that rely on high-energy environmental control installations. Even when circumstances do not demand it, agents usually insist on air-conditioning installations in order to achieve established market rental criteria.

With government departments and blue-chip companies occupying more and more air-conditioned space through the 1970s and 1980s, the development of new, naturally ventilated, office buildings was not considered to be commercially viable. Owner-occupiers involved in custom-build were not immune from this pressure, for they always ran the risk of being in the position of needing to assign (sub-lease) or sell their building on in a marketplace driven by institutional norms.

As other investment media such as equities became increasingly attractive in the early 1980s, it was clear that the peak of institutional investment in property had been reached. But, by then, institutional standards had established a grip that nobody seemed prepared to challenge (Cadman 1990). Their reification is clearly shown in the comments of this architect;

> there is a list of fifty things your building must have and these fifty things range from VAV (variable air volume) air-conditioning, 2.7 metre floor-to-floor ceiling heights, three lifts that allow thirty seconds travel or waiting time. All these things are absolutes as far as (agents) are concerned because they are selling it to their counterpart in another office and it's a circular thing.

The proliferation of highly specified office space had further consequences, generating dissatisfaction with the quality older buildings and leading some to argue that 'much of the city's existing stock is too tired, fragmented, and inadequate to meet new needs' (Duffy and Henney 1989). What had begun as an ideal-type standard for corporate headquarters had somehow turned into an industry norm, shaping development practice throughout the boom of the 1980s.[30] While high-profile users shared the desire for these highly specified super-offices the majority of occupiers were offered little choice. As demand outstripped supply, tenants represented little more than a taken for granted stream of income. A take it or leave it system operated in which rents were driven ever upward, encouraging new office developments to adopt the latest institutional specification in order to secure comparable rents. Copies of London's fully air-conditioned super-offices soon appeared in the Thames Valley, Bristol, Manchester and Edinburgh (Ferguson 1987).

During the late 1980s, the development process was controlled by a particular configuration of property interests. Developers had the upper hand and the details of the built environment reflect a particular approach to risk and investment (Guy and Harris 1997). In such a situation, energy efficiency was of marginal significance to speculative developers and owner-occupiers alike.

Nesting and using

In this section we describe a pathway of development that has quite different implications for the production of energy-efficient office buildings. The property market of mainland Europe has a different history and rates of owner-occupation are typically higher than in the UK. This has encouraged a tailor-made diversity of types and forms of office space. In the commercial sector, workers' representatives are often involved in designing the working environment, frequently insisting on adjustable heating systems and openable windows. Access to daylight tends to be controlled by more rigorous legislation (Burt 1992). This emphasis on user needs has a number of implications for property development.

First, there has been less stress on achieving the maximum rentable area, less weight placed on the production of prime space and less pressure to routinely achieve the 'highest' possible specification. For instance, a study of the French office market by Weatherall Green and Smith identified a common dislike of air-conditioning systems and of open plan offices, a cellular layout being generally favoured by French occupiers (Weatherall, Green and Smith 1989). Similarly, in Germany regulations on occupants' access to daylight mean that the 'depth' of buildings (from core to window) is typically shallower than in Britain. The result is a reduced rentable area and increased construction costs for the developer, but a more 'user-friendly' building of a type that appears to be widely desired (Williams 1990).

These are broad claims and conditions and practices obviously vary widely from one country to another. Nevertheless, across mainland Europe emphasis on the 'use' value of offices is in contrast to the 'Anglo Saxon' tendency to focus on 'exchange' value. As Duffy puts it, 'before the northern European office architect draws a single line, the users are already crowding around the drawing board' (Duffy 1989: 32).[31] The roots of this more user-centred style of property development are unclear. Some argue that it relates to the lack of a professional development culture as exists in the UK. For example, Vincent Renard has commented that 'France does not seem to have a tradition of private companies engaged in property development and management of the urban fabric as has long been the case in Britain' (Renard 1990: 160). Clearly, the role of the developer tends to be weaker in countries where people are inclined to rent their homes but buy their offices (Duffy 1991). Equally the lack of a developed investment market may favour custom build, or make it a cheaper option. Tax advantages and state funding often exist to support self-development, and legal structures tend to be codified in favour of the tenant rather than the landlord. This results in shorter, less binding, leases in which responsibility for repairs and maintenance does not lie with the tenant alone (Sweby Cowan Research 1992). Such arrangements give occupiers rather greater power and sustain a measure of flexibility and diversity in the form and specification of office buildings.

We now focus upon the recent history of commercial building in France in order to illustrate the specific determinants of real estate practices and to flesh out their implications for energy efficiency. France has a history of state intervention, both legal and financial, aimed at solving shortages of residential property and alleviating the difficulties of developing commercial property outside Paris (Erdmann 1992). Co-operative ownership is widespread,[32] and developments are often financed by a mixed collective of private, commercial and institutional investors (Stapylton-Smith 1994). Buildings have commonly been sold 'off-plan' to an end-user who would amortize the total cost over 15 years, effectively self-financing the development. French developers and investors rarely hold onto a building for long because they face high development taxes (around 20 per cent of development costs) if they do not sell on within five years. It is not just the procurement process that is at stake because differences in the legal, economic and cultural aspects of 'European' and 'Anglo-Saxon' real estate practices are just as pronounced. The lease terms of the rented sector seem generous by UK standards. Derived from post-war legislation, leases last for nine years with options to break in the third and sixth years in favour of the tenant. This creates a much greater sensitivity to market conditions. Tenants have the freedom to move as their needs change, particularly in conditions of oversupply. Rental increases are indexed and responsibilities for repair and maintenance are shared, the building's fabric and services generally being taken care of by the landlord. The legal obligations of landlords and tenants are enshrined in the Napoleonic codes that structure the

French legal system. This results in much shorter, standardized lease documents, which reduce the need for professional arbitration and that assume, and perhaps encourage, a less adversarial relationship between developer and occupier (Acosta and Renard 1993).

The greater emphasis on custom-build together with this more prescriptive legal arrangement has an impact on the structure of the real estate profession. A company requiring a new building may more readily organize finance, and approach an architect and commission exactly what it wants or can afford (Vale and Vale 1991). The role of the developer is here limited to the management of construction. Similarly, in the process of developing and letting space, the role of the agent is more often that of a broker than an adviser. Agents are rarely involved in the design process. Importantly, no single property actor dominates the development and letting process. The traditional French approach to real estate is in some ways a simpler affair than in the UK. The legal, commercial and professional structure is not designed to extract the maximum surplus economic value from the production of office space. Instead, there is a more equitable distribution of power between providers and users. As a result, the French occupier has the opportunity to actively shape standards of office space and performance.

The French do not seem to have wholly embraced the global office culture.[33] A survey by Catriona Allen showed that naturally ventilated offices in Paris are far more likely to be occupied by French than by 'international' users (Allen 1995). These environmental factors contribute to a Parisian preference for 'ill-suited' city centre refurbishments over the purpose built commercial offices at La Défence. Yet we should dispel the image of France as a 'green-office' utopia in which benign occupiers insist on appropriate spaces to the chagrin of greedy developers. While French office space seems to have avoided the worst excesses of the 'pursuit of the prime' as experienced in the UK, much of the stock is of poor quality. The thermal requirements of French building regulations are higher than those of the UK, but the insulation standards of much of the traditional French office stock are typically lower than those of modern British offices. Inefficient, inflexible and poorly constructed, these properties are the mirror-image of the over-specified British super-office. This is the source of a prevalent view that French real estate practices are years 'behind' those of the UK. Specifically, the local culture of office development and occupation is believed to inhibit technical change. In contrast, the drive towards international standards of space and performance, stimulated both by the requirements of trans-national corporations and the presence of international investors, is believed to act as a motor of change. But in what direction and with what consequences for energy efficiency?

Globalizing design

Between the mid-1980s and 1990 office rents doubled in Paris, leading to a massive rise in speculative development activity. Acosta and Renard (1993) suggest that while the present situation in Paris cannot yet be compared with that in cities such as Tokyo, where the price of land in central areas is now almost entirely disconnected from its use value, the character of the local office market has nonetheless shifted. Witness, for example, the rapid growth of the service sector, the expansion of information technology, and the internationalization of markets (Laing 1993). These shifts have affected the development desires of worldwide corporate players, the real estate expression of which takes the form of the 'intelligent office', uniting spatial flexibility, sophisticated servicing, and environmental efficiency. With the arrival of 'international tenants' and their global aspirations, a two-tier market has developed in France. There is a tangible struggle between the pull of global standards, supported by international users and investors, and the cultural preferences of local occupiers. The former, attracted to the purpose built and institutionally acceptable spaces of international zones, have pulled building structures up and out to accommodate more sophisticated services. Meanwhile the latter have clung to established centres, stretching existing buildings to the limit of their service and spatial capacity.

How then do we understand French real estate practice in relation to those of the UK? It is clear that the idea of a hierarchy in which the French are struggling to catch up with British standards is as flawed as the view that the French system represents an ideal model of user-oriented design. We simply see each set of practices as being different, each conditioned by their own cultural, commercial and legislative histories. In Paris we have identified an increasing tension between local real estate practices and those associated with global development and investment. A process of compromise may ensue in which international investors' zeal for global specifications is tempered by a more cautious population of local users. Resistance to the escalation, standardization and spread of 'prime', institutionalized space may prevent the spiralling of specifications that have characterized the British experience.

We have noted how the peculiar political and commercial conditions arising in the UK led to an escalation of standards out of all proportion to most user needs, also encouraging profligate patterns of energy consumption. The story of French property development takes a very different course. These differences do not simply reflect the environmental attitudes of individual property professionals. As we have suggested, the two forms of real estate practice have complicated, deeply rooted histories and traditions. In neither case is it right to identify occupants or owner-occupiers as the source of environmental action or to demonize speculative developers and investors. Both are implicated in sets of relationships and practices that, in their different ways, influence the construction of energy-efficient office space.

Fluctuating priorities

In this final section we track shifts in the British property market during the 1990s as a means of showing how property relationships and priorities change over time. In the early 1990s, vacancy rates in London were up to 15 per cent (the highest in Europe), while rental levels dropped by 20 per cent from the previous year (Klemann 1992). Although the market has now more or less recovered, there is still talk about whether the balance of power between investors and occupiers has really changed. Movements in the social organization of property development have stimulated debate about the future of 'realistically' specified office space.[34] As one developer commented:

> Change is happening now ... not just because people are becoming more environmentally aware, but because the development equation doesn't stack up at the moment. So it's a commercial business objective that's actually bringing about lower specification buildings ... but that's good if it also brings about environmentally sensitive buildings.

For a time, occupiers had the capacity to negotiate about the kinds of space they lease. In an effort to prevent rental income from sliding below economic levels, concessions, including shorter terms, break points and rent-free periods, were offered (Lizieri 1994). With rents dropping, service charges have became more visible (Owen 1992). This was significant for energy costs are normally the largest single element of these charges. As costs-in-use rose higher up the agenda (Harris 1993b), agents offered to analyse the cost of occupation as part of their consultancy service. Agents also recognized the commercial potential of efficient, non air-conditioned space (Smith 1993). With rental income declining, developers' priorities also changed. It was increasingly important to curb construction costs in order to make development equations stack up.[35] This prompted developers to re-assess conventional specifications inherited from the 1980s. Other cornerstones have also been overturned. There has been continuing debate about the need to revise 'upward-only rent reviews', 'confidentiality clauses' and the use of 'arbitration' in dispute resolution (McKibben 1993), all features said to introduce 'market distortions' artificially protecting the landlord's interest. Facilities managers have influenced the initial specification of buildings, co-ordinating service provision so as to maximize system efficiency (Owen 1992), and address the energy concerns of developers and occupiers (Melvin 1992). At the same time, growing public awareness of environmental issues has prompted companies to adopt 'green policies' or 'charters' and to seek a more 'back to basics' image (Parsa 1992). All this has had implications for property choices, and agents and developers are increasingly interested in adopting environmental friendliness as a letting and sales strategy (Barnard 1992).

These developments set a new scene in which energy efficiency figured not as an isolated running cost but as a potentially central consideration in the development process, perhaps also constituting a 'performance indicator of effective management' (Leaman 1992: 23). Properly handled, environmental innovation could be promoted as a 'way to achieve multiple benefits – for the developer, for the investors, for the occupants, and for the natural environment' (Wilson and Hedge 1992: xi).

Albeit for different reasons, developers and occupiers are now keen to consider 'actual' requirements in the design of office specifications. There is of course great reluctance to be the first to 'reduce' specification levels. A general conservatism masks vested interests in the preservation of existing practice. From designers whose fees are related to cost, to developers worried about market and legal risks involved in 'under specifying', resistance to change is significant. However, there are signs of a collective movement. The British Council of Offices, a diverse amalgam of interested property professionals, has produced a recommended 'appropriate' or 'realistic' specification for the 1990s.[36] The Building Services Research and Information Association (BISRIA) has produced an environmental code of practice for building services (Halliday 1996). Aimed at designers, surveyors, owners and occupiers these guidance documents promote the use of renewable resources. Higher standards of energy efficiency are an essential part of such a strategy.

'Less can do more' seemed to be the watchword for the 1990s developer (Lipton 1992: 58). On the other hand, most of the property professionals we interviewed felt that the wheel would turn again, that occupiers would be forced to accept institutional leases, and that investors would again wrest control from occupiers (South 1993). That is one scenario, but there are no certainties in the world of commercial property. It is just as plausible to argue that a new commitment to energy efficiency represents the first sign of a new partnership between occupiers and developers and a new form of real estate practice (Guy 1998). Having identified the benefits of energy efficiency, occupiers may stick to their guns, capitalizing on their new knowledge, conserving resources and limiting environmental damage. Similarly, developers might see lasting benefits in a continuing commitment to 'realistic' design, recognizing that this will help to cut construction costs, thereby stimulating development opportunities and providing more user-friendly offices. As one occupier commented;

> You can bring all this back to the environment if you want. The best way that environmental things work is that there is a cost attached. If we're looking at an office, here we have an asset under-utilized which is costing us a fortune. It costs us a fortune to buy it or to lease it and to run it, heat it, light it. Whether there are people here or not, you still have to do that. Yet we're not using it, so why have we got it? It's crazy.

This possibility takes us beyond the conceptual opposition of use and exchange value in which 'individuals and groups differ in which aspect (use or exchange) is most crucial to their own lives' (Logan and Molotch 1987: 2). Theoretically it also takes us beyond the economic realm in which the 'strategies of agents' are framed in terms of 'rent relations' alone (Luithlen 1994: 125).

Our stories of difference and change in the French and British property markets have shown how the 'motives, operations and financial stability of different types of developer' mark them as distinct (Adams 1995: 531). We have also recognized the competing objectives that 'shape the role of landowners, investors, agents and occupiers in the development process' (Adams 1994: 90–131). In this way, we have been able to follow the dynamic struggle over the form and specification of future office space as shaped by local tensions between investors, developers and occupiers, each operating in an increasingly global context.

In France, the future of energy-efficient office development depends on the balance between the demand for 'better quality' space (as defined by international users and global financiers) and local traditions, tastes and practices, which appear to resist the imposition of 'prime' but energy-intensive specifications. In the UK, the future of energy-efficient office development appears to depend upon a loosening of the institutional grip on real estate practices. Again, this is not to suggest that moving from an investors' to a tenants' market automatically produces more energy-efficient buildings. Just as we must recognize that investors' influence is contingent on particular markets, so we must understand that occupiers' priorities are structured by changing needs and concerns, some of which are shaped by agents and others involved in the property business.

Either way, it is clear that the model of a split market in which owner-occupiers respond to calls for energy efficiency, and speculative developers resist them, fails to capture the dynamics of office development or to recognize the inter-dependence of the interests at play. We have argued that movements in the collective balance of power between occupiers and developers create different opportunities for energy-efficient design, and that there tend to be more such chances when occupiers have the upper hand. But that is not the same as suggesting that owner-occupiers are more likely to put the lessons of energy-related science into practice than speculative developers in any one market situation. Instead, we have argued that both are caught within a more macro narrative of property development and that this is the level at which the pathways of energy-efficient office design are shaped and formed.

8 Conclusions

The last three chapters have focused on the ways in which energy efficiency is embedded in the world of practice. Following the interests and actions of manufacturers, designers, developers, investors, and a host of other players, we have shown how the issue of energy efficiency is wrapped up with, and shaped by, social, political and commercial processes. In so doing, we have pointed to a way of seeing energy efficiency, which is quite unlike the techno-economic perspective we considered in earlier chapters. The remaining tasks for this book are threefold: to review the paradigms of energy efficiency identified in the preceding chapters; to unpack the relationship between competing ways of seeing energy efficiency; and to outline an agenda for energy-related research and policy that recognizes the socially constructed nature of its subject, and that promises to identify socially as well as technically viable strategies and responses.

Challenging taken-for-granted assumptions about the dynamics of technical choice and change, our aim has been to broaden debate about energy research and policy. Our analysis has taken us beyond simple evaluations of the efficacy of technical knowledge or the peculiarities of consumer behaviour, allowing us to offer more complex explanations for the success and failure of energy-saving initiatives. In developing a broadly sociological approach we have had to consider the roles, responsibilities and strategies of a much wider range of factors than that which typically populates studies of energy efficiency. This has meant delving more deeply into the changing contexts that frame energy-related knowledge and choice. In this way, we have shown how competing conceptions of energy efficiency generate very different practices and standards of energy performance.

Understanding energy efficiency

More a sociology of knowledge than of consumer behaviour, this book began by looking at the evolution of the energy problem from the point of view of research managers and others involved in funding and promoting energy-related building science. Borrowing ideas from research and science policy and tapping into the experiences and practices of research managers in the USA, UK, Ireland, Finland, Sweden and France, Chapter 2 considered the shaping of technical research in six countries. We asked a number of questions. What kinds of problems did the energy crisis throw up, how were these defined and addressed, what new knowledge was thought to be required and how was such expertise supposed to relate to building practice? There were good reasons for expecting research programmes to differ given the climates, cultures and histories of construction that characterize the six countries selected for study. Despite striking differences in the social organization of science in Finland, Sweden, Ireland, France, the UK and the USA, the anticipated role of technical enquiry was much the same. Time and again, building scientists and their funders subscribed to a remarkably uniform, remarkably linear, model of research, development, demonstration and dissemination. Chapter 3 considered the practical and theoretical implications of methodological convergence. A shared belief in the transferability of building science and widespread acceptance of associated theories of technical change meant that resources were devoted to remarkably similar activities in each country. Universally, demonstration projects were built and monitored and instances of 'best practice' recorded. As if blinded by the material similarity of building structures, researchers viewed the 'non-technical' determinants of energy performance as obstacles and barriers to technological progress.

In the first part of the book we showed how energy efficiency has been defined as a subject of technical enquiry and illustrated the model of technology transfer that underpins the design and management of government-funded programmes of research and development. In constructing and promoting research programmes, policy makers, energy analysts, engineers and technologists drew on a range of conventional theories about how the social world works, and about how social actors can best be influenced. The second part of the book examined these conceptual models in greater detail, starting with ideas about the process of technological diffusion.

The blend of physical, political, and even climatic determinism that characterizes techno-economic explanations of technological diffusion is quite distinctive. Chapter 5 got to grips with some of these elements of received wisdom, exploring them with reference to the uneven history of mineral wool insulation. This might have seemed a strange point of entry. After all, insulation is such a simple device that it barely qualifies as a 'technology'. Although insulation represents one of the most effective means of reducing energy consumption in buildings, it has been used in different ways in neighbouring and climatically similar countries. In recounting the narratives of diffusion in France, the UK and Denmark, we took a critical look at explanations usually offered for different standards of insulation and for the uneven adoption of energy-saving measures.

Accepted wisdom suggests that levels of insulation depend upon climate, technical potential, economic circumstance and regulatory policy. The actions, interests and structure of the industry play no part in this explanatory framework. However, we identified important cultural differences in the way this crucial business is organized. Participating in a close community, Sweden's insulation industry has opportunities to influence events in ways that would be unimaginable for those operating within the fragmented British market. French insulation manufacturers have a stronger presence and greater influence than their British counterparts, again because of the specific features of their commercial history. Seen this way, the material stuff of energy conservation depends as much upon these commercial characteristics as it does upon regulatory systems or climatic conditions.

Reflecting on the individualistic theories of action that inform so many energy-saving initiatives, Chapter 6 examined the responses of 'decision-makers' confronting apparently similar technical choices about housing design. Energy-saving campaigns routinely address Local Authorities, Housing Associations and private sector builders as interchangeable producers of a common commodity. Comparison of the distinctive pressures and opportunities that characterize each of these worlds was enough to make us think again about the social structuring of energy-related choice. In Britain, Housing Associations, experiencing greater organizational freedom coupled with declining state aid, are fighting to maintain their traditional standards of construction. Meanwhile, Local Authorities are struggling with a depreciating stock, reduced improvement grants and widening social problems. Private builders are typically keen to cut costs and generally resist efforts to upgrade energy standards. Looking across these contrasting organizational worlds, it is clear that apparently similar technical decisions are coloured by very different assessments of economic and political risk. This study led us to conclude that designers' practices are much more strongly determined by the contexts in which they operate than by their personal knowledge or individual enthusiasm for energy efficiency.

The techno-economic perspective suggests that the market penetration of well-established or successfully demonstrated technologies is frequently impeded by non-technical barriers. These include such complicating features as the landlord–tenant relationship (in which those who might invest in energy-saving technologies do not directly benefit from lower fuel bills) or the nature of the professional fee structure (if professional fees represent a percentage of the cost of heating and ventilating systems, engineers have no incentive to design systems with less energy-intensive equipment). The array of potential obstacles is impressive, as demonstrated by those who have sought to itemize barriers to energy-efficient technology transfer (Evans 1991) or to draw distinctions between 'structural and behavioural barriers' (Hirst 1992). Repeated calls for 'additional research to understand barriers, to assess their importance, sector by sector and to examine the effectiveness of policy options that might overcome them' (Hirst and Brown 1990: 278; Bondi 1988) reinforce the belief, first, that such barriers are real and, second, that governments have a legitimate part to play in supporting efforts to correct these and other market imperfections.

Our investigation of the design and specification of commercial office building in France and the UK suggested that this representation of 'barriers' depends upon an unwarranted conceptual separation of the social and the technical. Although the barrier vocabulary points observers in the direction of the social world, it does so in a way that promises to obscure as much as it reveals. In emphasizing differences between landlords and tenants, and between owner occupied and speculative developments, energy analysts tend to overlook the dynamics of the property market and the fluctuating priorities of those involved. Rather than viewing energy-saving technology as a stream of potential that is variously blocked by confounding features of the social world, we used the story of office development to highlight the co-evolution of social and technical systems of energy efficiency. We showed how specifications for office developments tend to be driven by the investment criteria of institutional funders. Conditions of over-supply, together with a legal framework favourable to the suppliers and owners of buildings, have encouraged specifications to escalate way beyond the 'needs' of occupiers. As a consequence, excessive air-conditioning and elaborate systems of energy management have become the norm in the UK. Comparison with other European real estate practices suggests that tenant-led markets permit more efficient levels of specification. Although a clash of interest between landlord and tenants, or suppliers and users may well impede energy efficiency in some circumstances, this is not universally the case. As we showed, interests in energy efficiency depend upon the social organization of property development and this differs between cultures and over time. All three cases – the study of the insulation industry in Denmark, France and Britain; the comparison of public and private sector housing in Britain, and the review of office development in France and Britain – challenged conventional theories of technical change and introduced alternative models. In the process they highlighted a number of puzzles about the relationship between standardized, seemingly transferable technical expertise and the day-to-day realities of building practice. Critically, each chapter questioned the flow of knowledge between research and practice or, to put it more abstractly, each showed how the local knowledge of practitioners interacted with the more 'cosmopolitan' or universal forms of knowledge produced by building scientists and researchers.

Technical convergence, cultural diversity

Looking back, we have told an uneven story of similarity and difference: we find different research cultures operating with the same borrowed logic of building science and technical change. We find similar building types located within profoundly different social contexts and we find that the conclusions of building science are expected to take root in wildly different contexts of action. How can we understand these tensions and what are their consequences for energy research and practice?

In seeking to understand these tensions we have developed an 'hourglass model' of technical convergence and cultural diversity. This image helps make sense of the way in which differences in the cultures of knowledge production are reduced through scientific and methodological consensus, and how diverse research environments generate seemingly transferable technical conclusions. At this point the complexity of practice comes into play again as variable contexts of action confound the ready application of technically proven strategies.

As the hourglass figure suggests, the unifying language of building science cuts across national variety at the level of research systems. Differing relationships between government, industry, and academia are pushed into the background by standardized methods and shared theories of technical potential and technology transfer. Common commitment to a techno-economic paradigm goes hand-in-hand with

CULTURAL VARIETY IN THE PRODUCTION OF KNOWLEDGE

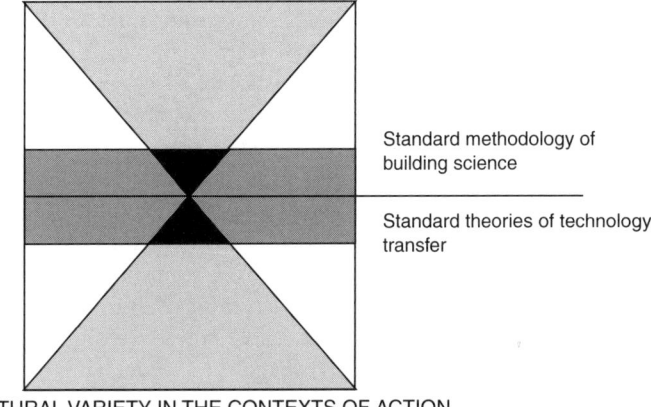

Standard methodology of
building science

Standard theories of technology
transfer

CULTURAL VARIETY IN THE CONTEXTS OF ACTION

Figure 9 Hourglass diagram

a shared belief that the sequential stages of research, development, demonstration and dissemination flow from top to bottom of the diagram. Agreement about the nature of technical change informs the development of matching programmes of technology transfer. As conventionally understood, these initiatives are designed to inform and persuade populations of socially anonymous and often bewilderingly irrational individuals (situated in the bottom half of the diagram) on whom the application of building science ultimately depends.

Looking up from the bottom of the hourglass, we offer a different interpretation of the appropriation of technical expertise and the application of energy-saving knowledge. As our analysis of the contexts of decision-making and the realities of building practice suggest, accepted theories of technical change fail to take account of the routine complexities of social action. Our three case studies show that technical change is a social process and that more or less energy-efficient choices are made in response to changing opportunities and pressures. It is not simply a question of transferring technologies upon people. Instead, knowledgeable actors creatively adopt and adapt strategies and practices that suit their changing circumstances. Sometimes these favour energy efficiency, sometimes not.

The effect of the hourglass shape, its waist narrowed by techno-economic conventions, is such that knowledge and ideas can only pass from top to bottom when squeezed together and funnelled through this slim but obligatory passage point. Alternative interpretations of technical change are sidelined and, as a result, much of the innovation process goes unexplained. Excluded from the research agenda, the world(s) of practice reappear as obstacles and barriers. Homogenizing the problem of innovation to make it tractable in scientific terms, energy-efficiency researchers take little heed of practitioners' and consumers' own knowledge and little note of how their actions are enmeshed in various systems, structures and established conventions. The challenge of energy efficiency is consequently misconstrued as challenge of 'getting the message across', regardless of the situational pressures and priorities that frame technical choices on the ground. Driven by the 'need' to establish standardized recommendations and solutions, technical researchers struggle to understand how interests and practices can vary so widely over space and time. From this perspective it is hard to recognize how practitioners' capacities for action evolve, or to comprehend the very different organizational constraints and opportunities that determine the production of otherwise similar building types.

The significance of these omissions becomes clear in the setting of research priorities. Research investment is repeatedly and almost exclusively dedicated to the definition and resolution of technical problems, including the problem of technology transfer. Much less effort is invested in understanding the dynamics of social, political and commercial change that frame the decisions and practices in which energy efficiency is embedded.

Revising the hourglass figure just a little, we now put it to another purpose. Figure 10 plots the traditional boundaries of social and natural scientific expertise.

As we have discovered, building science occupies the central band of the hourglass, patrolling and directing interaction between the worlds of research and practice. The social sciences, in the form of human dimensions research, have a thin slice of this action, helping to market energy efficiency and understand and overcome non-technical barriers. We have, however, identified other forms of social enquiry, positioned above and below the central band of energy-related research and development. For instance, many sociologists of science have devoted themselves to studying both the production and the diffusion of knowledge (Cozzens *et al.* 1989; Gibbons *et al.* 1994). Equally, we have identified sociological research that examines the histories and practicalities of design, technology and energy consumption in the built environment. Again, there are many good examples to which we can refer (Kempton and Montgomery 1982; Janda 1996;

CULTURAL VARIETY IN THE PRODUCTION OF KNOWLEDGE

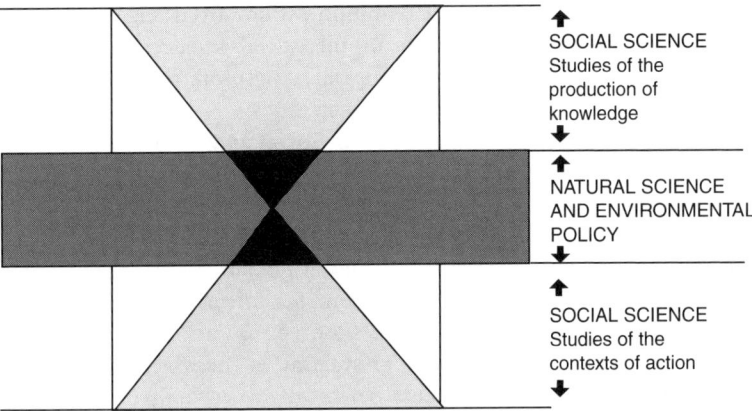

↑
SOCIAL SCIENCE
Studies of the
production of
knowledge
↓

↑
NATURAL SCIENCE
AND ENVIRONMENTAL
POLICY
↓

↑
SOCIAL SCIENCE
Studies of the
contexts of action
↓

CULTURAL VARIETY IN THE CONTEXTS OF ACTION

Figure 10 Hourglass diagram modified

Wilhite *et al.* 1996; Lutzenhiser 1997; Cooper 1998). The key question concerns the relationship between socially and historically sensitive analyses, and the narrow theories of technical progress and change that currently dominate the central zone. What are the chances of re-building research and policy around more context-sensitive theories of the relationship between knowledge and practice? Would this involve widening the waist of the hourglass or inventing another model altogether? In the next section we consider new roles for social science and the potential for developing and working with less restrictive models of technical choice and change.

Reconstructing research

The analysis we have presented in this book has practical implications for energy-related research and policy. At the simplest level, we argue that greater attention should be paid to the changing contexts of energy-related decision-making. This means recognizing that proven technologies might work perfectly well across a range of technically similar situations, but that they are not necessarily transferable from one social environment to another. As the case studies illustrate, better understanding of the social organization of choice allows us to identify circumstances that currently conspire to favour energy efficiency. If decisions are as socially structured as we suggest, policy makers and research funders should revise their theories of human action and associated 'hypodermic' models of communication. By doing so, and by drawing upon relevant research in the social sciences, they might devise and adopt contextually appropriate and more precisely tailored strategies for promoting energy efficiency.

It is easy to prescribe strong doses of social analysis but we have to allow that established approaches have a history and a logic of their own. Opportunities to modify the technical research agenda and re-shape theories of technology transfer depend on the current organization of environmental research and development. As we have already explained, national research environments differ. Social scientists may, for example, form part of a 'close community' (as in Sweden, Ireland or France) or contribute to systems of 'networking expertise' (as in France). Both arrangements permit their active involvement. By comparison, the research cultures of 'contracting knowledge' (exemplified by the UK) and, to a lesser extent, 'co-ordinated contracting' (as in the USA), restrict the terms of interdisciplinary interaction. Ideas do move between the central zone of technical research (as illustrated in the hourglass figure) and the surrounding margins of social science, but they do so in ways that are themselves socially structured. This is important because nationally specific forms of research management have immediate consequences for social scientists seeking to engage with environment-related policy makers. If the patterns described here apply elsewhere, we might expect the social sciences to have less influence on energy-efficiency research and policy in Britain and the USA than in countries like France and Sweden. Likewise, the chances of developing more robust socio-technical theories and policies appear to be rather greater where expertise is networked or where communities are close.

If social scientists are to make a substantial contribution to the analysis and promotion of energy efficiency, it will be necessary to go way beyond the 'human dimensions' of what remain essentially technological programmes. This implies re-structuring research funding and re-defining agendas. Such moves are vital if there is to be any hope of learning about where the most promising opportunities for energy-saving action really lie, or of going with the grain of current practice in order to exploit such opportunities. If the community of relevant experts was expanded to include a variety of social scientists (such as those whose ideas we considered in Chapter 4), we might expect to see a corresponding shift in working definitions of the energy problem. Greater appreciation of the social structuring of technical change would, for instance, transform the view of energy efficiency as a largely technical issue and encourage a more contextual understanding of innovation. Greater sensitivity to the dynamics of practice would complicate the abstraction of generalized knowledge and lead to a thickening of the waist at the centre of the hourglass. New ways of linking the worlds of research and practice might follow, perhaps drawing upon rather than ignoring changing patterns of tacit knowledge and practice.

The analysis of building science and practice we have presented here is itself an example of the more extensive contribution that social science might make to the analysis and promotion of energy efficiency. At this point it is useful to spell out the key features of what has this far been a somewhat covert approach. In developing an alternative understanding of energy-saving action we have abandoned the search for 'human factors' that might speed or impede technological diffusion. In conducting interviews we did not try to assess respondents' attitudes to energy or the environment, or identify motivational or financial 'triggers' that might convince them to adopt energy-efficient measures. We did not stand back and, from a safe but critical distance, review, evaluate or measure the 'real' potential for energy efficiency in the sectors we examined. Nor did we take energy efficiency to be a matter of social construction alone. Instead, we sought to develop a broadly sociological approach to the everyday practicalities of energy-efficient design. In the process we made use of a range of intellectual resources, tapping into a somewhat varied repertoire of ideas about research, technology, and energy-related practice. By drawing these analytical resources together we have, however, developed a rather precise narrative about the production of scientific knowledge and technical change in the fields of building design and development.

Features of this approach are mirrored in the methods we adopted. We began by talking with researchers and policy-makers about how they understood the problem of energy efficiency and about how their work might help to reduce energy demand. These lines of questioning revealed the parameters of the techno-economic paradigm and allowed us to learn about energy efficiency from the perspective of those involved in technical research. In designing the case studies described in Chapters 5, 6 and 7 we steered a different course. Rather than concentrating on energy efficiency as a subject in its own right, we followed the routines and agendas of practitioners designing and developing the built environment. This sometimes involved explicit discussion of technical strategies and energy-saving measures, but more often than not such themes were buried within a more wide ranging discussion of social and commercial practices. In order to understand the implicit structuring of energy efficiency we had to abandon pre-conceived ideas about the dynamics of environmental change or technology transfer. Instead, we had to listen and learn about the changing pressures and priorities that governed the strategies of different actors. Following energy through the design and development process we discovered how technical strategies mesh with commercial tactics, and how the plans of one practitioner often clashed with the designs of another.

Understanding the day-to-day organization of building and design is not a task that can be undertaken at a distance. Methodologically, we had no option but to establish a dialogue with practitioners, compare their contexts of action and identify the dynamic aspects of context and circumstance that structured their expectations and practices. Throughout, we have been interested in how practitioners view themselves, and how they tell their tales of practice. As a result, much is missing from our story. For example, we have not considered government's role in making energy-related policies, nor have we paid much attention to the changing price of fuel, the privatization of the utilities, or the evolution of environmental politics. We could have given much more explicit consideration to the power of national and global economies in setting local development and investment priorities. We might have pointed to regional differences, or to the characteristics of new buildings as compared with renovation projects. There are other stories to be told and other studies to be done. We have, however, demonstrated the relevance and value of an approach that escapes the confines of the human dimensions agenda and that directly engages with the localized knowledge of practitioners operating in constantly changing professional environments. This book has been as much about developing such an approach as about reaching substantive conclusions.

In the process, it has helped to define an alternative way of seeing energy and with that, a fresh set of research questions. Having re-cast the problem in broadly sociological terms, there are many possible points of departure. For example, we clearly need to know more about how tacit knowledge of energy efficiency develops. Leaving questions of technology transfer behind, the real challenge is to map the growth and development of ideas about energy efficiency in the multiple worlds of building design and practice. What are the changing frameworks of understanding within which energy efficiency is constructed as a problem or a solution? How do building codes, design conventions and manufacturing systems become established, when and in what ways do they shift, and how do these processes frame the definition of socially as well as technically viable opportunities for energy-saving action?

As we have seen, many energy-related actions are governed by practices and priorities that have nothing to do with energy efficiency *per se*. Taking this point to heart, we might highlight the energy implications of contemporary efforts to solve the housing crisis, to promote the 24-hour city or to develop the paperless office. Taking this logic further, we might decide that it is a mistake to focus research efforts on energy efficiency at all. Recognizing the extent to which such issues are embedded within the conventions of building design and the development of the urban environment, we might ask more radical questions about what energy is actually used for. What services, activities and lifestyles does energy consumption make possible? In a process of reverse translation, we might begin to enquire about the patterns of production and consumption that lie behind the profile of escalating demand for energy. This might lead us to think about how expectations of comfort evolve and the extent to which they are, for example, the cause or the consequence of the rapid spread of air-conditioning around the world. What part do researchers and designers play in creating as well as responding to consumers' needs and expectations of the built environment?

Read as an analysis of the relationship between environmental research and policy, on the one hand, and practical environmental action, on the other, this book has explored familiar tensions between local and global interests, between social and natural science, and between cultural variety and seemingly transferable technical solutions. Extending the reach of our conclusions still further, we have, we hope, made the case for a wide-ranging sociology of the environment: one that engages with such ordinary issues as energy, waste and water, and one that does justice to the social ordering of choices, problems and practices.

Notes

1 http://www.wuppertal-institut.de/Sites/divisions.html
2 These images formed part of the UK Energy Efficiency Office's 1991 campaign.
3 See, for example, the UK Building Research Establishment's Energy and Environmental Assessment Model (BREEAM) and the development of databases cataloguing the green credentials of building materials.
4 In the US, the Energy Information Administration of the Department of Energy collects and analyses data on energy consumption in different building sectors, and for different states.
5 The Finnish LINKKI programme, now in its second phase, addresses the human and social/community dimensions of energy efficiency.
6 Energy research came to a virtual standstill during the Reagan years.
7 The Department of Energy's office of building technology, state and community programmes supports research and development in human factors with the aim of addressing the societal opportunities and barriers to implementing energy efficiency in buildings (US Department of Energy 2000). This agenda concentrates on consumers' purchasing patterns, and the information available to key decision-makers.
8 Even so, this has been the subject of an EU-funded study, the results of which led to the conclusion that passive solar energy provides around 14.5 per cent of space heating and that the figure could be increased to 15.5 per cent by the year 2010 (Burton, Doggart and Crompton 1989).
9 The ETSU house design study programme was itself designed as a form of action research in which 'real' designers worked with solar energy experts and cost advisors to explore ways of improving standard house types. The design study team was then charged with the task of making sense of this work and extracting the lessons learned.
10 In this subtractive approach, estimates of solar energy depend on estimates of incidental gains, that is on estimates of the nature and character of average activity within an average household (Bartholomew 1994). Despite their central role in the representation of solar energy, calculations of average activity are fraught with uncertainty.

11 CADDET's objective is to broaden and improve the collection and exchange of information on energy-efficient technologies in the end-use sector. By this means it aims to provide governments, utilities, industrial concerns and other end-users with a better understanding of such technologies, thereby encouraging both more informed decision-making and improved replication of successful demonstrations (Abel *et al.* 1992: v).

12 For example, since 1995 the UK Government has provided over £60 million funding for the Energy Savings Trust to promote the efficient use of energy in the UK. See: http://www.est.org.uk/est/achievements/925481037ach.html.

13 As the Association for the Conservation of Energy points out, the UK government's earlier ambition to improve the efficiency of energy use by 20 percent between 1983 and 1990 has not been achieved: only a 9 per cent improvement was recorded between 1983 and 1990 (Environment Committee 1993b: 39).

14 Rogers' 1962 analysis of innovation suggests that new ideas are first adopted by a limited number of risk-taking 'early adopters'. Their experience inspires confidence in the bulk of the population, but it takes time before the most reluctant section of the population, 'the laggards', are convinced. Hence the graph of diffusion represents an 'S' curve.

15 Stern and Aronson (1984) identify five archetypes of energy user; the 'investor', 'consumer', 'member of a social group', the 'expresser of personal values', and 'avoider of problems', (See Janda 1998: 30).

16 In picking our way through these resources we do not distinguish between the different histories of the sociology of science and the sociology of technology (see Pinch and Bijker 1989), nor do we pay much attention to the intellectual ancestries of the concepts we borrow.

17 Two buildings, standing side by side, may well share all sorts of physical and material characteristics. Investigation of their respective modes of production and consumption might nonetheless reveal profoundly different interests, logics and rationales, which might in turn explain differences in recorded patterns of energy consumption.

18 EURISOL, the UK's mineral wool trade association, claims that each square metre of 5-mm thick insulation reduces carbon dioxide emissions by about 1 tonne over the lifetime (50 years) of a building (EURISOL 1990).

19 Cavity walls consist of two layers of brick or block work with a gap between. Cavity wall insulation involves blowing mineral fibre or polystyrene bead, or injecting foam, into this gap.

20 Badly insulated walls can become damp and it is extremely difficult to remove foam once it has set.

21 For example, with respect to the revision of Part L (the thermal part) of the building regulations (Raman and Shove 2000).

22 See, for example, http://www.energy-efficiency.gov.uk/

23 By 1989/90, 53 per cent of new Housing Association homes were

5 per cent below Parker Morris standards. Two years later the figure had slipped to 68 per cent (Karn and Sheridan 1994).

24 Parker Morris standards were critical. The main priorities for improvement were in floor space and heating – floor space to accommodate all the activities undertaken in a typical household and heating to ensure that the space could be used at all times of the year, in the winter as well as the summer (Goodchild and Furbey 1986: 80).

25 See *New Builder*, 5th March 1993.

26 For example, see: Beazer Engineering Services (1993) 'Proposed amendments to part L of the building regulations: a review and cost analysis', Unpublished Report, Somerset: Beazer Homes (West) Ltd.

27 See for example, *Energy Efficiency in Office Buildings: An Initial Market Research Study*, Property Market Analysis for BRECSU, 1988.

28 Custom-built office development makes up less than 10 per cent of the UK market.

29 See Building Research Establishment (BRE),Energy Consumption Guide 19.

30 See the discussion in 'H.Q. Buildings', *CSW – The Property Week*, 21st October 1993: 27–39.

31 Peter Ambrose has written about the rarity of speculative development in Sweden (at least pre-1991) and about how 'alternative modes of property development and management' permit 'striking examples of user control' (Ambrose 1994: 194).

32 The system of 'co-propriété' (co-ownership) which has existed since about 1938 has encouraged this trend.

33 Here it is important to stress the importance of Paris as the commercial centre. Paris is the headquarters location of 96 per cent of banks and 70 per cent of insurance companies. 60 per cent of all French office space is concentrated in Paris (Erdmann 1992). Accordingly Paris also possesses a huge rental sector (estimated at over 40 million m^2 of commercial space).

34 See the PROCORD business index, reviewed in *The Times* 20 October 1993: 33.

35 Construction cost analysis in the early 1990s showed that a basic air-conditioned building cost between £893 and £1,706/m^2, compared with a basic, non-air-conditioned building at between £622 and £904/m^2 (Davis *et al.* 1991).

36 See, for example, *Best Practice in the Specification of Offices* (1997) British Council for Offices, Reading.

Bibliography

Abel, E., Aronson, S., Jagemar, L., Nilsson, P. (1992) *Learning from Experiences with Energy Efficient Retrofitting of Office Buildings*, Sittard: CADDET.

Abel, E., Andersson, J., Berglund, B., Bylin, G., Dawidowicz, D., Hanssen, S., Lindvall, T., Lundgren, B., Norlen, U., and Samuelson, I. (1999) 'The healthy building: status of research', Stockholm: Swedish Council for Building Research.

Acosta, R. and Renard, V. (1993) *Urban Land and Property Markets in France*, London: UCL Press.

Adams, D. (1994) *Urban Planning and the Development Process*, London: UCL Press.

Adams, D. (1995) 'The British commercial development industry', *European Planning Studies* 3:(4): 531–42.

ADEME (1993) *Promoting Sustainable Development*, Paris: ADEME.

Akrich, M. (1992) 'The description of technical objects', in W. Bijker and J. Law (eds) *Shaping Technology/Building Society*, Cambridge, MA: MIT Press.

Allen, C. (1995) 'A study of office environments in central Paris', MA Thesis, University of Newcastle.

Ambrose, D. (1994) *Urban Process and Power*, London: Routledge.

Audits of Great Britain, Home Audit Division (1985) 'Awareness of and attitudes towards cavity wall insulation', Unpublished Report for the Building Research Establishment, Garston, Watford.

Ball, M. (1983) *Housing Policy and Economic Power*, London: Methuen.

Ball, M. (1988) R*ebuilding Construction: Economic Change in the British Construction Industry*, London: Routledge.

Barnard, S. (1992) 'A breath of fresh air', *Estate Times* Sept. 18: 41–2.

Bartholomew, D. (1994) 'Occupancy data', Unpublished Report for the Building Research Establishment, Garston, Watford.

Baum, A. (1991) *Property Investment Depreciation and Obsolescence*, London: Routledge.

Beck, U. (1992) *Risk Society: Towards a New Modernity*, London: Sage

Beck, U. (1995) *Ecological Politics in an Age of Risk*, Cambridge: Polity Press.

Bell, M. (1998) *An Invitation to Environmental Sociology*, Thousand Oaks, CA: Pine Forge Press.

Bijker, W. (1992) 'The social construction of fluorescent lighting or how an artefact was invented in its diffusion stage', in W. Bijker and J. Law (eds) *Shaping Technologies, Building Society*, Cambridge, MA: MIT Press.

Bijker, W.E. (1995) 'Sociohistorical technology studies', in S. Jasanoff, G. Markle, J. Peterson and T. Pinch (eds) *Handbook of Science and Technology Studies*, Cambridge, MA: MIT Press.

Bijker, W.E. and Law, J. (eds) (1992) *Shaping Technology/Building Society*, Cambridge, MA: MIT Press.

Bijker, W.E., Hughes, T.P. and Pinch, T. (eds) (1987) *The Social Construction of Technological Systems*, Cambridge, MA: MIT Press.

Bondi, H. (1988) *Evaluation of the Research and Development Programme in the Field of Non-Nuclear Energy 1985–1988,* Research Evaluation Report No.24 EUR 11834 EN/1, Brussels: Commission of the European Communities.

Boss, A., Orbalsi, A. and Sutcliffe, S. (1993) *House Design Studies Overview*, Energy Technology Support Unit S1362, Harwell: ETSU.

Bromley, M. (1993) 'Getting credit where credit's due', *Housing*, April: 8.

Brown, M. and White, D. (1992) 'Evaluation of Bonneville's 1988 and 1989 Residential Weatherization Program: a northwest study of program dynamics', Oak Ridge, CA: Oak Ridge National Laboratory, ORNL/CON-323.

Brunskill, R. and Clifton Taylor, A. (1977) *English Brickwork*, London: Ward Lock.

Burningham, K. and Cooper, G. (1998) 'Misconstructing constructionism: a defence of the political utility of social constructionist approaches to environmental problems', in A. Gijswijt, F. Buttel, P. Dickens, R. Dunlap, A. Mol and G. Spaargaren (eds) *Sociological Theory and the Environment, Proceedings of the Second Woudschoten Conference*, Amsterdam: SISWO, University of Amsterdam.

Burt, A. (1992) 'A comparison of UK/USA and European business parks', *Property Management* 9(3): 17–9.

Burton, S., Doggart, J. and Crompton, A. (1989) 'The performance and general success of passive solar buildings throughout Europe', in T. Steemers and W. Palz (eds) *Science and Technology at the Service of Architecture: 2nd European Conference on Architecture 4–8 December, Paris.* Dordrecht: Kluwer Academic Press.

CADDET (Centre for the Analysis and Dissemination of Demonstrated Energy Technologies) (1998) 'Energy efficiency', Sittard: NOVEM.

Cadman, D. (1984) 'Property finance in the UK in the post-war era', *Land Development Studies* 1: 61–82.

Cadman, D. (1990) 'The environment and the urban property market', *Town and Country Planning* 59(10): 267–70.

Cadman, D. and Catalano, A. (1983) *Property Development in the United Kingdom – Evolution and Change*, Reading: College of Estate Management.

Callon, M. (1987) 'Society in the making' in T. Bijker, T.P. Hughes and T. Pinch (eds) *The Social Construction of Technical Systems*, Cambridge, MA: MIT Press.

Callon, M., Laredo, P. and Rabehariosa, V. (1992) 'The management and evaluation of technological programs and the dynamics of techno-economic networks: the case of AFME', *Research Policy* 21: 215–36.

Chadderton, D.V. (1991) *Building Services Engineering*, London: E&FN Spon.

Charfas, J. (1991) 'Skeptics and visionaries examine energy savings', *Science* 251: 154–6.

Chaudhary, V. (1993) 'Housing group warn of steep rent increases for tenants, *The Guardian*, 23rd August: 7.

Chevin, D. (1991) 'Wet through', *Building* 13th December: 50–1.

Clowes, C. (1993) *Scheme Development Standards B Draft Proposal*, London: Housing Corporation.

Cockram, A. and Arnold, P. (1984) 'Urea formaldehyde foam cavity wall insulation: reducing formaldehyde vapour in dwellings', Building Research Establishment Information Paper 7/84, Garston, Watford: BRE.

Colclough, J. (1965) *The Construction Industry in Britain*, London: Butterworth.

Colquhoun, I. and Fauset, P. (1991) *Housing Design in Practice*, London: Longman Scientific and Technical.

Command 2250 (1993) *Realising Our Potential: A Strategy for Science, Engineering and Technology*, London: HMSO.

Command 4814. (1971) *The Organisation and Management of Government R&D: A Framework for Government Research and Development*, London: HMSO.

Cooper, G. (1998) *Air Conditioning America: Engineers and the Controlled Environment 1900–1960*, Baltimore: Johns Hopkins University Press.

Cottrell, F. (1955) *Energy and Society: The Relation between Energy, Social Change and Economic Development*, New York: McGraw Hill.

Courtney, R. (1997) 'BRE- past, present and future' *Building Research and Information: the International Journal of Research, Development and Demonstration* 25(5): 285–92

Cowan, R.S. (1983) *More Work for Mother: the Ironies of Household Technology from the Open Hearth to the Microwave*, New York: Basic Books.

Cowan, R.S (1987) 'How the refrigerator got its hum', in D. McKenzie and J. Wajcman (eds) *The Social Shaping of Technology*, Milton Keynes: Open University Press.

Cozzens, S., Healey, P., Rip, A., and Ziman, J. (eds) (1989) *The Research System in Transition*, Dordrecht: Kluwer Academic Publishers.

Danish Energy Agency. (1992) 'Energy efficiency in Denmark', Copenhagen: Danish Energy Agency.

Dard, P. (1986) *Quand l'énergie se domestique . . . Observations sur dix ans D'expériences et D'innovations Thermiques dans L'habitat*, Paris: CSTB, Plan Construction et Architecture.

Davis, Langdon and Everest (eds) (1991) *Spon's Architectural and Builders Price Book* London: Chapman and Hall.

Department of the Environment (1991) *English House Condition Survey: Supplementary Energy Report*, London: HMSO.

Department of the Environment (1993) *Climate Change: Our National Programme for CO2 Emissions*, London: HMSO.

Department of the Environment (1994a) *Energy Efficiency in Buildings*, Cm 2453, London: HMSO.

Department of the Environment (1994b) *Energy Efficiency in Council Housing*, London: HMSO.

Diamond, R. (1984) 'Energy use among the low-income elderly: a closer look' *Proceedings, 1984 ACEEE Summer Study on Energy Efficiency and Buildings*, Washington, DC: ACEEE F52–66.

Disco. C., Rip, A., and van der Meulen, B. (1992) 'Technical innovation and the universities: divisions of labour in cosmopolitan and technical regimes', *Social Science Information* 31(3): 465–501.

Duffy, F. (1989) 'The European challenge', *Architects Journal* 17th August: 32–43.

Duffy, F. (1991) 'Vive la difference', *Europroperty* October: 29–33.

Duffy, F. and Henney, A. (1989) The changing city, London: Bulstrode Press.

Eder, K. (1993) *The New Politics of Class: Social Movements and Cultural Dynamics in Advanced Societies*, London: Sage.

Egan, C., Kempton, W., Eide, A., Lord, D., and Payne, C. (1996) 'How customers interpret and use comparative graphics of their energy use' *Proceedings, Human Dimensions of Energy Consumption 1996 ACEEE Summer Study*, Washington, DC: ACEEE: 189–99.

Egan, J. (1998) *Rethinking Construction: The Report of the Construction Taskforce*, London: Department of the Environment, Transport and the Regions.

Energy Efficiency Office (1991), 'Insulating your home, no.2', Dd8240826 ENGY JO546NJ, London: HMSO.

Environment Committee (1993a) *Fourth Report, Energy Efficiency in Buildings*, vol. 1, p. 648-I, London: HMSO.

Environment Committee (1993b) *Fourth Report, Energy Efficiency in Buildings*, vol. 2, p. 648-II, London: HMSO.

Erdmann, E. (1992) *Property,* London: Mercury Books.

EURIMA (1990) 'Thermal insulation means environmental protection', Brussels: EURIMA.

EURIMA (1991) 'Thermal insulation standards in housing in Europe', Brussels: EURIMA.

EURISOL (1981) 'Thermal insulation of cavity walls', Redbourn: EURISOL.

EURISOL (1990) 'Pollution reduction through energy conservation', Redbourn: EURSOL.

Evans, R. (1991) *Barriers to Energy Efficiency: A Report to the Energy Efficiency Office*, London: Energy Efficiency Office.

Farhar, B. (1993) 'Trends in public perceptions and preferences on energy and environmental policy: executive summary', Golden, CO: National Renewable Energy Laboratory NREL/TP-461-4857a.

Fedeski, M. (1991) 'The environmental performance of buildings: design aid for architects', PhD Thesis, Welsh School of Architecture, University of Wales College of Cardiff.

Fedeski, M. (1993), 'Design aid to suit design practice' in N. Foster and H. Scheer (eds) *Proceedings, Solar Energy in Architecture and Urban Planning, Florence 17–21 May 1993*, Bedford: H.S. Stephens and Associates.

Ferguson, A. (1987) 'Offices for professionals', *Estates Gazette*293, 12th September: 1226–8.

Friends of the Earth (1990) *How Green Is Britain?* London: Hutchinson Radius.

Friends of the Earth (1994) *The Climate Resolution*, January, London: Friends of the Earth.

Gibbons, M., Limoges, C., Nowotny, H., Schwartzman, S., Scott, P., and Trow, M. (1994) *The New Production of Knowledge: The Dynamics of Science and Research in Contemporary Societies*, London: Sage.

Gicquel, R. and Cools, C. (1989) 'The CEC Project PASSYS: achievements in the period 1986–1989' in T. Steemers and W. Palz (eds) *Science and Technology at the Service of Architecture: 2nd European Conference on Architecture 4–8 December, Paris*. Dordrecht: Kluwer Academic Press.

Gillard, M. and Tomkinson, M. (1980) *Nothing to Declare: the Political Corruptions of John Poulson*, Sheffield: Platform Books.

Giovannini, B. and Baranzini, A. (eds) (1998) *Energy Modelling: Beyond Economics and Technology*, Geneva: Centre for Energy Studies, University of Geneva.

Goobey, R. (1992) *Bricks and Mortals*, London: Century Press.

Goodchild, B. and Furbey, R. (1986) 'Standards in housing design: a review of the main changes since Parker Morris', *Land Development Studies* 3: 79–99.

Graham, I. (1992) 'Housing association consortium development', *Journal of Property Finance* 3(4): 485–91.

Groak, S. (1992) *The Idea of Building: Thought and Action in the Design and Production of Buildings*, London: E&FN Spon.

Grubb, M. (1991) *Energy Policies and the Greenhouse Effect*, Aldershot: Dartmouth.

Grubb, M. (1992) *Emerging Energy Technologies: Impacts and Policy Implications*, London: The Royal Institute of International Affairs.

Guy, S. (1998) 'Developing alternatives: energy, offices and the environment', *International Journal of Urban and Regional Research* 22(2): 264–82.

Guy, S. (1999) 'Evil developers and green fairies', in B. Fairweather S. Elworthy, M. Stroh and P. Stephens (eds) *Environmental Futures*, London: Macmillan.

Guy, S. (2000) 'Framing environmental choices: mediating the environment in the property business', in S. Fineman (ed.) *The Business of Greening*, London: Routledge.

Guy, S. and Harris, R (1997) 'Property in a risk society: towards marketing research', *Urban Studies* 34(1), January: 125–40.

Haas, P. (1990) *Saving the Mediterranean: the Politics of International Environmental Cooperation*, New York: Columbia University Press.

Haas, P. (1992) 'Introduction: epistemic communities and international policy coordination', *International Organisation* 46(1): 1–35.

Halliday, S. (1996) *Environmental Code of Practice for Building and their Services - Case Studies*, vol. 1, April. London: The Building Services Research and Information Association (BSRIA).

Hannigan, J. (1995) *Environmental Sociology: a Social Constructionist Perspective*, London: Routledge.

Hanson, N. (1981) 'Observation as theory laden' in brown, in S. Fauvel and R. Finnegan (eds) *Conceptions of Inquiry*, Milton Keynes, UK: Open University Press.

Harris, A. (1983) 'Radical economics and natural resources', *International Journal of Environmental Studies* 21: 45–53.

Harris, D. (1993a) 'How to cut office energy costs', *The Times*, 20th October: 33.

Harris, D. (1993b) 'Continental edge to competition', *The Times*, 20th October: 33.

Harvey, D. (1989) *The Urban Experience*, Oxford: Blackwell.

Hedges, A. (1991) *Attitudes to Energy Conservation in the Home B Report on a Qualitative Study*, London: HMSO.

Hensen, J. and Hand, J. (1993) 'Use of sophisticated building energy simulation tools', in N. Foster and H. Scheer (eds) *Proceedings, Solar Energy in Architecture and Urban Planning, Florence 17–21 May 1993*, Bedford: H.S. Stephens and Associates.

Hilmo, T. (1990) 'Review of barriers to a better environment', *Geografiska* 72 B: 124.

Hinchliffe, S. (1995) 'Missing culture: energy efficiency and lost causes', *Energy Policy* 23(1): 93–5.

Hirst, E. (1992) 'Making energy efficiency happen', in M. Kuliasha, A. Zucker and K. Ballew (eds) *Technologies for a Greenhouse Constrained Society*, Michigan: Lewis Publishers.

Hirst, E. and Brown, M. (1990) 'Closing the efficiency gap: barriers to the efficient use of energy', *Resources, Conservation and Recycling* 3: 276–81.

Howard, N. (1994) 'Materials and energy flows', in *Cities, Sustainability and the Construction Industry*, Swindon: Engineering and Physical Science Research Council: 11–14.

Hughes, T.P. (1983) *Networks of Power: Electrification in Western Society, 1880–1930*, Baltimore: Johns Hopkins University Press.

Hughes, T.P. (1988) 'The seamless web: technology, science, etcetera, etcetera', in B. Elliott (ed.) *Technology and Social Process*, Edinburgh: Edinburgh University Press.

Hutcheon, N.B. and Handegord, G. (1983) *Building Science for a Cold Climate*, John Sussex: Wiley & Sons.

International Energy Agency (1991), *Energy Policies of IEA Countries*, Paris: IEA

Jaffe, A. and Stavins, N. (1994) 'The energy paradox and diffusion of conservation technology', *Resource and Energy Economics* 16(2): 91–122.

Janda, K. (1996) 'Designing from experience: the effects of research on practice', in *Commercial buildings: Program Evaluation, Proceedings, 1996 American Council for an Energy Efficient Summer Study*, Washington, DC: ACEEE Press

Janda, K (1998) 'Building change: effects of professional culture and organisational context on adopting energy efficiency in buildings', PhD Thesis, University of California, Berkeley, CA.

Jonas, P.J. (1981) 'Energy conservation and energy demand and supply in the UK', in R. Derricot and S. Chissick (eds) *Energy Conservation and Thermal Insulation*, Sussex: John Wiley and Sons, pp. 97–108.

Karn, V. and Sheridan, L. (1994) *New Homes in the 1990s: a Study of Space and Amenity in Housing Sector and Private Sector Production*, York: Joseph Rowntree.

Kealy, L. (1989a) 'Ideology and information in low energy design', in T. Steemers and W. Palz (eds) *Science and Technology at the Service of Architecture: 2nd European Conference on Architecture 4–8 December, Paris*. Dordrecht: Kluwer Academic Press.

Kealy, L. (1989b) 'Case studies in passive solar design: bringing the good news or comforting the faithful', in T. Steemers and W. Palz (eds) *Proceedings, Science and Technology at the Service of Architecture: 2nd European Conference on Architecture 4–8 December, Paris.* Dordrecht: Kluwer Academic Press.

Kelly, R. (1994) 'Beyond the limits of logic', *The Guardian*, 1st October: 21.

Kempton, W. (1993) 'Will public environmental concern lead to action on global warming?', *Annual Review of Energy and the Environment* 18: 217–43.

Kempton, W. and Montgomery, L. (1982) 'Folk quantification of energy', *Energy* 7: 817–27.

Kempton, W., Boster, J., and Hartley, J. (1995) *Environmental Values in American Culture*, Cambridge, MA: MIT Press.

Kennedy, J. (1993) 'What housing policy?', *Architecture Today* 37, April: 20–8.

Kleeman, W. (1992) 'Global office market survey', *Facilities* 10(10): 5.

Kuhn T.S. (1962) *The Structure of Scientific Revolutions*, Chicago: The University of Chicago Press.

Laing, A. (1993) 'Changing business: post Fordism and the work-place', in F. Duffy, A. Laing and V. Crisp (eds) *The Responsible Workplace*, London: Butterworth Architecture.

Langstaff, M. (1992) 'Housing associations: a move to centre stage', in J. Bichall (ed.) *Housing Policy in the 1990s*, London: Routledge.

Lash, S., Szerszynski, B. and Wynne, B. (eds) (1995) *Risk, Environment and Modernity: Towards a New Ecology*, London: Sage.

Latour, B. (1987) *Science in Action*, Milton Keynes: Open University Press.

Latour, B. and Woolgar, S. (1986) *Laboratory Life: the Construction of Scientific Facts,* Princeton: Princeton University Press.

Lawrence Berkeley Laboratory (2000) *From the Lab to the Marketplace: Making America's Buildings More Energy Efficient*, Berkeley, CA: Lawrence Berkeley Laboratory. Online. Available HYPERLINK http://eetd.lbl.gov/CBS/Lab2Mkt/TOC.

Lea, F. (1971) *Science and Building: A History of the Building Research Station*, London: HMSO.

Leach, G. (1991) 'Policies to reduce energy use and carbon emissions in the UK', *Energy Policy* December: 918–25.

Leaman, A. (1992) 'Complexity in buildings', *Facilities* 10(9): 23.

Leeden, B. (1975) 'Letter to the *Architects Journal*', *Architects Journal* 4 June 1975.

Lewis, J.O., and O'Toole, S. (1990) 'EC support for architects', in C. den Ouden (ed.) *Proceedings of a Workshop on Design Support to Architects, Edinburgh 19–20 June 1990,* Dordrecht: EGM Engineering.

Lipton, S. (1992) 'On offices', *CSW B The Property Week* 29th Oct: 58.

Lizieri, C. (1991) 'The property market in a changing world economy', *Journal of Property Valuation and Investment* 11 September: 201–13.

Lizieri, C. (1994) 'Property ownership, leasehold forms and industrial change', in M. Ball, A.C. Pratt (eds) *Industrial Property*, London: Routledge.

Löfstedt, R. (1992) 'Lay perspectives concerning global climate change in Sweden', *Energy and the Environment* 3(2): 161–75.

Logan, J. and Molotch, H. (1987) *Urban Fortunes: The Political Economy of Place*, California: University of California Press.

Lovins, A. (1992) *Energy Efficient Buildings: Barriers and Opportunities*, Boulder, CO: E-Source.

Luithlen, L. (1994) *Office Development and Capital Accumulation in the UK*, London: Avebury.

Lutzenhiser, L. (1992) 'A cultural model of household energy consumption', *Energy B the International Journal* 17: 47–60.

Lutzenhiser, L. (1993) 'Social and behavioural aspects of energy use', *Annual Review of Energy and Environment* 18: 247–89.

Lutzenhiser, L. (1994) 'Sociology, energy and interdisciplinary environmental science', *American Sociologist* Spring: 58–79.

Lutzenhiser, L. (1997) 'Social structure, culture and technology: modelling the driving forces of household energy consumption', in P. Stern, T. Dietz, V. Ruttan, R. Socolow and J. Sweeney (eds) *Consumption and the Environment: the Human Causes*, Washington, DC: National Academy Press.

Lutzenhiser, L. and Shove, E. (1999) 'Contracting knowledge: the organizational limits to interdisciplinary energy research and development in the US and the UK', *Energy Policy* 27(4): 217–27.

Macey, S. and Brown, M (1990) 'Demonstrations as a policy instrument with energy technology examples', *Knowledge, Creation, Diffusion, Utilization* 11(3): 219–36.

Madanipour, A. (1996) *Design of Urban Space: An Inquiry into a Socio-spatial Process*, Sussex: Wiley.

Malpass, P., and Means, R. (eds) (1993) *Implementing Housing Policy*, Buckingham: Open University Press.

Marriot, O. (1967) *The Property Boom*, London: Hamish Hamilton.

Martell, L. (1994) *Ecology and Society: an Introduction*, Cambridge: Polity Press.

Marvin, H. and Mackinder, M. (1985) 'Information and experience in architectural design', Research Paper 23, Institute of Advanced Architectural Studies, York: University of York.

McElroy, L. (1993) 'The energy design advisory service as an aid to a new working frame', in N. Foster and H. Scheer (eds) *Proceedings, Solar Energy in Architecture and Urban Planning, Florence 17–21 May 1993*, Bedford: H.S. Stephens and Associates.

McIntosh, A. and Sykes, S. (1985) *A Guide to Institutional Property Investment*, London: Macmillan.

McKenzie, D. and Wajcman, J. (eds) (1985) *The Social Shaping of Technology*, Milton Keynes: Open University Press.

McKibben, E. (1993) 'Commercial leases – further changes', *Estates Gazette* Nov. 6: 109–10.

Melvin, J. (1992) 'Speculative development and facilities management', *Facilities* 10(9): 18–22.

Mulkay, M. (1979) *Science and the Sociology of Knowledge*, London: George Allen and Unwin.

National Audit Office (1994) *Buildings and the Environment*, London: HMSO.

National Federation of Housing Associations (1985) *A Brief Introduction to Housing Associations*, Unpublished Report, London: National Federation of Housing Associations.

National Federation of Housing Associations (1988) 'Warm homes: a manifesto', Unpublished Report, London: National Federation of Housing Associations.

(NCIA) National Cavity Insulation Association (1989) 'An update on cavity wall insulation: an opportunity to reduce pollution, energy costs and state subsidies', Haslemere: NCIA.

Nelson, S. (1993) 'A study of the use of design-build contracts by housing associations', *Housing Review* 42(4): 69–70.

Newby, H. (1991) 'One world, two cultures: sociology and the environment' British Sociological Association, *Network* 50: 1–8.

Nielson, J.H. (1990) 'Denmark's energy future', *Energy Policy* January/February: 82.

Nye, D.E. (1998) *Consuming Power: A Social History of American Energies*, Cambridge, MA: MIT Press.

Olson, R. (1988) *Energy in the Built Environment*, Stockholm: Swedish Council of Building Research.

Owen, D. (1992) 'The facilities manager and the speculative developer', *Facilities* 10(9): 8.

Parsa, A. (1992) *The Impact of Environmental Issues on Commercial Property*, London: Royal Institution of Chartered Surveyors.

Pinch, T. and Bijker, W.T. (1989) 'The social construction of facts and artifacts: or how the sociology of science and the sociology of technology might benefit each other', in W.E. Bijker, T. Hughes and T. Pinch (eds) *The Social Construction of Technological Systems*, Cambridge, MA: MIT Press.

Plender, J. (1982) *That's the Way the Money Goes*, London: Andre Deutsch.

Price, D.J. de Solla (1962) *Little Science, Big Science*, New York: Columbia University Press.

Raman, S. and Shove, E. (2000) 'The business of regulation', in S. Fineman (ed.) *The Business of Greening*, London: Routledge.

Ravetz., A. (1980) *Remaking Cities*, London: Croom Helm.

Rayner, S. (1991) 'A cultural perspective on the structure and implementation of global environmental agreements', *Evaluation Review* 15(1): 75–103.

Redclift, M. (1996) *Wasted: Counting the Costs of Global Consumption*, London: Earthscan.

Redclift, M. and Woodgate, G. (1994) 'Sociology and the environment: discordant discourse', in M. Redclift and T. Benton (eds) *Social Theory and the Global Environment*, London: Routledge

Renard, V. (1990) 'Land and property development in France', in P. Healey and R. Nabarro (eds) *Land and Property Development in a Changing Context*, London: Gower.

Rip, A. (1992) 'Expert advice and pragmatic rationality', in N. Stehr and R. Ericson (eds) *The Culture and Power of Knowledge: Enquiries into Contemporary Societies*, Berlin: Walter de Gruyter.

Rogers, E.M. (1962) *Diffusion of Innovation*, New York: Free Press.

Rosa, G., Mahlis, G., and Keating, K. (1988) 'Energy and society', *Annual Review of Sociology* 14:149–72.

Rydin, Y. (1986) *Housing Land Policy*, London: Gower.

Schipper, L. (1987) 'Energy conservation policies in the OECD: did they make a difference?', *Energy Policy* December: 538–48.

Schipper, L., Meyers, S., and Kelly, H. (1985) *Coming in From the Cold: Energy-wise Housing in Sweden*, Washington, DC: Seven Locks Press.

Schwartz, H. (1996) 'The rise and decline of energy bureaucracies', *Proceedings, Human Dimensions of Energy Consumption 1996 ACEEE Summer Study*, Washington, DC: ACEEE: 189–99.

Seaden, G. (1997) 'The future of national construction research organizations', *Building Research and Information: the International Journal of Research, Development and Demonstration* 25(5): 250–7.

Shackley, S. and Wynne, B. (1996) 'Representing uncertainty in global climate change science: boundary-ordering devices and authority' *Science Technology and Human Values* 21(3): 275–302.

Shove, E. (1995) 'Constructing regulation and regulating construction', in T. Gray (ed.) *Environmental Politics in the 1990s*, London: Macmillan.

Shove, E. (1997a) 'Contracting knowledge: commissioned research and the sociology of the environment', in C. Samson and N. South (eds) *The Social Construction of Social Policy*, London: Macmillan.

Shove, E. (1997b) 'Energy knowledges', in *Proceedings of the European Council for an Energy Efficient Economy 1997 Summer Study, Sustainable Energy Opportunities for a Greater Europe*, Copenhagen: Danish Energy Agency/ECEEE.

Shove, E. (1998a) 'Gaps, barriers and conceptual chasms: theories of technology transfer and energy in buildings', *Energy Policy* 26: 1105–12.

Shove, E. (1998b) 'Relevance, independence and capture', *Building Research and Information: the International Journal of Research, Development and Demonstration* 2(6): 386–9.

Shove, E. and Raman, S. (1996) 'Big stick or bendy stick? Regulating for energy efficiency' *Proceedings of the 1996 ACEEE Summer Study on Energy Efficiency in Buildings*, Washington, DC: ACEEE Press.

Shove, E. and Wilhite, H. (1999) 'Energy policy: what it forgot and what it might yet recognise', in *Energy Efficiency and CO2 Reduction: the Dimensions of the Social Challenge*, Proceedings of the 1999 Summer Study, European Council for an Energy Efficient Economy, Paris: ADEME Editions.

Shove, E., Lutzenhiser, L., Guy, S., Hackett, B., and Wilhite, H. (1998) 'Energy and social systems', in S. Rayner and E. Malone (eds) *Human Choice and Climate Change, vol. 2, Resources and Technology*, Ohio: Battelle Press.

Smith, R. (1993) 'The need for a realistic specification', *Specification of Buildings, Proceedings 2nd Annual Conference*, Reading: British Council of Offices.

Socolow, R. (ed.) (1978) *Saving Energy in the Home: Princeton's Experiments at Twin Rivers*, Cambridge, MA: Ballinger Press.

South, G. (1993). 'David Hunter B a man of action', *Chartered Surveyor Weekly*, 30th Sept: 20.

Stapylton-Smith, D. (1994) 'Investing in centres in copropriété, *Estates Gazette* 11 June: 111–13.

Stearn, J. (1994). 'Space B the final frontier', *Inside Housing* 22nd July: 10–1.

Stern, P.C. and Aronson, E. (eds) (1984) *Energy Use: The Human Dimension*, New York, NY: W.H. Freeman and Company.

Stewart, A. (1993). 'Housing Association methods condemned', *Housing*, 29 January: 5.

Sweby Cowan Research Services (1992) *The Guide to European Property Investment*, vol. 1, Waterlow, London.

Thurber, J. (1963) *Vintage Thurber*, London: Hamish Hamilton.

Tombazis, A. and Preuss, S. (1999) 'The Avax office building, Athens', in J.O. Lewis and J. Goulding (eds) (1999) *European Directory of Sustainable and Energy Efficient Building*, London: James and James.

Trudgill, S. (1990) *Barriers to a Better Environment*, London: Belhaven Press.

Tyler, M. (1991) 'Sick buildings: carrying the can', *Architects Journal* 21st–28th August, 194: 48–50.

US Department of Energy (2000) 'R and D for human factors and community systems', Online. http://www.eren.doe.gov/buildings/research_human.html.

Vale, B. and Vale, R. (1991) *Towards a Green Architecture London*, London: RIBA Publications.

Vale, B. and Vale, R. (1997) 'The autonomous house, a model for suburban sustainability', Paper presented at the Catalyst '97 Conference, 5–8

December, Canberra. Online HTTP: http//www.gaia.org/secretariats/genoceania/documents/autnmshse.html.

Van de Perre, R. (1993) 'Improving confidence in computer simulation results', in N. Foster and H. Scheer (eds) *Proceedings, Solar Energy in Architecture and Urban Planning, Florence 17–21 May 1993*, Bedford: H.S. Stephens and Associates.

Vine, E. (1995) 'International DSM and DSM program evaluation: an INDEEP assessment' *Proceedings of the 1995 ECEEE Summer Study: The Energy Efficiency Challenge for Europe*, Stockholm: ECEEE.

Vine, E. Payne, C., and Weiner, R. (1993) 'Comparing the results of energy efficiency programs: the creation of a national data base on energy efficiency programs (DEEP)' LBL Report 33655, Berkeley, CA: Lawrence Berkeley Laboratory.

Volti, R. (1992) *Society and Technological Change*, New York: St Martins Press.

Wallace, D. (1995) *Environmental Policy and Industrial Innovation*, London: Earthscan Publications.

Warkov, S. and Meyer, J. (1993) *Solar Diffusion and Public Incentives*, Lexington, MA.: DC Heath.

Weatherall, Green & Smith and Grosvenor Estate Holdings (1989) *European Property Market Handbook B France*, Paris and London: Weatherall Green & Smith Research.

Webster, A (1991) *Science, Technology and Society*, London: Macmillan.

Wilhite, H. and Wilk, R. (1987) 'A method for self-recording household energy-use behavior', *Energy and Buildings* 10: 73–9.

Wilhite, H., Nakagami, H., Masuda, T., Yamaga, Y. and Haneda, H. (1996) 'A cross cultural analysis of household energy-use behaviour in Japan and Norway' *Energy Policy* 24(9): 795–803.

Wilk, R. (1996) *Economies and Cultures: Foundations of an Economic Anthropology*, Oxford: Westview Press.

Williams, V. (1990) *The Occupier's View: Business Space in the 1990s*, London: Building Use Studies.

Wilson, S. and Hedge, A. (1992) *The Office Environment Study*, London: Building Use Studies.

Wouters, P. and Vandaele, L. (1993) 'The CEC Project PASSYS: from research to practice', in N. Foster and H. Scheer (eds) *Proceedings, Solar Energy in Architecture and Urban Planning, Florence 17–21 May 1993*, Bedford: H.S. Stephens and Associates.

Yearley, S. (1991) *The Green Case*, London: Harper Collins.

Yearley, S. (1996) *Sociology Environmentalism Globalization*, London: Sage.

Index